Land Matters

Power Struggles in Rural Ireland

ETHEL CROWLEY

The Lilliput Press • Dublin

ACKNOWLEDGMENTS

This book is one of the products of many years of research on rural Ireland. It has been aided by inputs from friends and colleagues, but I would especially like to thank the many interviewees who have been so generous in sharing information about their lives with me.

I would also like to thank Marsha Swan and Antony Farrell at The Lilliput Press for their highly professional editorial work.

My partner, Jim MacLaughlin, has shared in all of the ups and downs along the way. His contribution has been unique and immeasurable.

First published 2006 by
The Lilliput Press
62–63 Sitric Road, Arbour Hill
Dublin 7, Ireland
www.lilliputpress.ie

Copyright © Ethel Crowley, 2006

ISBN 1 84351 081 2

A CIP record for this title is available from The British Library.

10 9 8 7 6 5 4 3 2 1

Set in 10.5 pt on 13.5 pt Sabon
Printed in Dublin by ColourBooks, Baldoyle

Contents

ABBREVIATIONS

ACOT	An Chómhairle Oiliúna Talmhaíochta
AFT	An Foras Talúntais
BSE	bovine spongiform encephalopathy
CAP	Common Agricultural Policy
CIWF	Compassion in World Farming
CSF	Cork Social Forum
DAF	Department of Agriculture and Food
DoE	Department of the Environment
EAGGF	European Agriculture Guarantee and Guidance Fund
EEC/EC/EU	European Economic Community/European Community/ European Union
EHO	Environmental Health Officer
EPA	Environmental Protection Agency
ESA	Environmentally Sensitive Area
FFI	Family Farm Income
FMS	Farm Modernisation Scheme
GATT	General Agreement on Tariffs and Trade
ICA	Irish Countrywomen's Association
ICMSA	Irish Creamery Milk Suppliers Association
IFA	Irish Farmers' Association
IMF	International Monetary Fund
IOFGA	Irish Organic Farmers and Growers Association
IPC	Integrated Pollution Control
KIO	Keep Ireland Open
LETS	Local Exchange Trading System
NAFTA	North American Free Trade Agreement
NATO	North Atlantic Treaty Organization
NHA	Natural Heritage Area
NSS	National Spatial Strategy
REPS	Rural Environment Protection Scheme
RRI	Rural Resettlement Ireland
RSPB	Royal Society for the Protection of Birds
SAC	Special Area of Conservation
SD	sustainable development
SFP	Single Farm Payment
SMA	Supplementary Measure A
SME	small and medium enterprises
SPA	Special Protection Area
TNC	trans-national corporation
UKROFS	UK Register of Organic Food Standards
UN	United Nations
UNESCO	United Nations Educational, Scientific & Cultural Organization
WTO	World Trade Organization

Foreword

Saturday has come to be a hectic day on our Atlantic side of the hill, loud with straining tractors and lorries revving at the bends. Home from a week of building other people's houses, the young men are finishing their own: a ribbon of construction sites winds out towards the mountain. What's left of the weekend goes to the sheep, the drains, the walls their grandfathers built and REPS requires to be kept in repair. It's a landscape in a strange flux, emptier than ever of livestock, or people walking the fields, but filling up with houses, many of which will be empty most of the year. The nearest village is doubling in size, becoming at last the town it always pretended to be, but one largely for second homes and retirement. Waves of change press headlong from the east – from Dublin, from Europe – but there is little sense that Irish people know what kind of countryside they want, or what pattern of farming life they should foster or preserve.

In this brave and significant book, Ethel Crowley investigates the social and psychological changes in rural Ireland since accession to the EEC. She exposes the gross inequalities that flowed from a Common Agricultural Policy based on intensive production. With no less radical vigour, she picks apart the social implications of the new 'green capitalism', whose managerial ethos leaves so many small-farm communities feeling marginalized and powerless. Her close study of the Rural Environment Protection Scheme (REPS) is enriched by many kitchen conversations and by interviews that explore the 'mysterious territory between EU principles and Irish practice'. She brings a sociological scalpel to policies, structures and directives in a way that lays them open to the light: for sheer readable information and reference, her book is invaluable.

It is its radicalism, however, that makes it memorable. One is not used, in Ireland, to sociology-with-attitude, let alone such directness of language. It goes behind the buzzword phrase of 'sustainable development' to challenge the comfort this continues to give to capitalist agriculture, increase in farm size, and intensification of land use and pollution. Crowley finds a future for small-farm Ireland in the organic movement and the bioregionalism of farmers' markets and locally branded foods. Her case-study exploration of West Cork, with its 'neo-tribes' of settlers and cultural frictions, ends with an optimistic faith in the social innovations such conflict can produce. This idealism and conviction, along with a sociologist's discipline, empower a remarkably timely book.

Michael Viney
Thallabawn, Co. Mayo

Introduction

'Farming', he told me, 'is like ironing underpants – a pointless exercise.' Sometimes the best sociological insights are gained on barstools. In one sentence, this young man had captured the pessimism and despondency felt by many farming men and women in contemporary Ireland. They feel frustrated at the constantly changing EU[1] rules that govern their activities, the spiralling inflation rate and low prices they acquire for their products, the heavy bureaucracy they encounter at every turn, but perhaps more than all of these, the increasingly negative perception of farming by the public at large.

The announcement on 29 September 2004 that Mary Coughlan was to replace Joe Walsh as Minister for Agriculture coincided with the National Ploughing Championships, the Mecca of rural Ireland. On the national airwaves, one woman made a heartfelt plea from that venue that the new minister would do all she could to 'improve understanding' of the difficulties faced by farmers. While a small cadre of these have profited hugely from EU monies, the vast majority of farm families have to supplement their incomes with other jobs. They have to make do with commodity prices that in many cases are proportionately lower than they were twenty years ago, while supermarket chains keep them in a stranglehold. Many of those

I

who cannot sustain their families in these conditions lease their land to other farmers, or perhaps sell up altogether. Large parts of the countryside have now been zoned for roads or housing, and environmentalists' concerns about protection of landscape and heritage are beginning to be taken seriously in official circles. In this competitive and aggressive climate, one needs to be articulate and organized in order to be taken seriously and have one's views represented in public. Some rural 'claimsmakers' have this ability, while others do not.

This book is an investigation of how this situation came about. After the War of Independence, small farmers were publicly portrayed in Ireland as the bedrock of society, the epitome of the values that made the emerging small nation distinctive and valuable. How did we get to the point that we are at now, when that same group is often seen as a throwback to a past that we would rather forget? What have been the processes that produced this change, and who are the main groups or individuals responsible? This book provides:

- an explanation of the crucial ideas that shape and drive EU and Irish rural policy
- an outline of the main policy decisions and their social impacts
- the most important inputs of key individuals and groups
- an example of one rural region and its attempts to surf the waves of globalization

Lots of stories have been told about rural Ireland. Some of the socially acceptable ones have become stitched into the cultural fabric, while others have been buried, never again to be heard. The stories of the women, of the emigrants, of the children in industrial schools, of the bachelor farmers, of the Travellers, had years to wait before they found a public voice. I hope to convince readers of the richness and potency of a sociological perspective, to enhance and challenge our views and deepen our social understanding.

Sociology examines and analyses the connections between different people's everyday actions and the factors that influence the contexts in which these actions occur. People's choices are circumscribed and constrained by factors like social class, race and gender, which have different meanings in different contexts. Karl Marx, one of sociology's founding fathers, said that people can create their own history, but not in conditions of their own choosing.[2] In order to analyse human actions, sociologists need to show sensitivity to cultural, geographical and historical contexts. When studying a phenomenon like the survival strategies of Irish farmers, one needs to clarify the precise social, cultural and economic factors that affect it, at this time and in this place. Here is a fictional example:

Paul, a young Irish farmer, is very aware of the problems he will face farming in Ireland in the near future. He decides to raise ostriches on his farm because he saw them on a study trip to New Zealand with a focus group from the local branch of the Irish Farmers' Association (IFA). He was exposed to this influence largely because of his class background. He comes from a sizeable farm where he can afford to take such risks, and his parents encourage him to use the knowledge he gained at agricultural college, and to innovate where he can. He therefore has the self-confidence and financial resources to keep up with global economic trends and to try new things. This will naturally have a positive bearing upon his prospects for long-term survival on the land.

Paul's neighbour, Seán, on the other hand, is not blessed with the same resources. His father died young, he left school early, his farm is small, he missed agricultural college and is struggling financially. He wouldn't be able to afford to go on the trip to New Zealand, even if he were an IFA member. His father was not a member, and neither is he. Anyway, he wouldn't dare take a chance on a new venture like ostriches, as his margins just wouldn't stretch that far. Instead, he uses the local pub as a place to meet contacts, through whom he gets a part-time job on a construction site nearby. The ready cash is needed to pay the bills. He is too tired when he comes home in the evening to do any more than the basics on the farm – just enough to keep things ticking over. If he lost his labouring job, he would have to lease his land to Paul, who always needs more land to expand his production and try out new ventures.

Paul's farm is highly integrated into the global economy, as he is heavily dependent upon EU cattle and milk prices. These have dropped dramatically in recent years due to overproduction throughout Europe. Seán's farm is also dependent upon EU support, but he has signed up for an agri-environmental scheme. In the medium term, the amount that Paul receives can only drop, and the small amount Seán receives will remain steady. Paul could use his personal resources to overcome his problems, however, while Seán is more likely to be stuck in a downward economic spiral and dependent on casual employment.

Sociology can come to the rescue here, to analyse the past, present and even future for the thousands of Pauls and Seáns in rural Ireland. Seamus Heaney's poem 'Digging' is a reflection on the onerous farmwork done by his father's generation. In the poem, he says that he will not follow in his father's footsteps, but instead:

> Between my finger and my thumb
> The squat pen rests
> I'll dig with it.

Digging with a pen could indeed be seen as a metaphor for the work of the sociologist. Digging to find the roots of human behaviour is an apt description of sociology, the science of society. Sociologists view human beings not just as individuals, but as social beings who are at once the products and creators of societies. Societies are composed of families, communities, institutions, miscellaneous social groups (from religious cults to political parties), and social structures. Some of these are visible to the ordinary observer; to see others requires a more specialized training. Digging to this extent is by definition a subversive activity, because it subverts our common sense. It is constantly trying to find the hidden story behind the official version that dominates public discourse on particular issues. When presented with information, the sociologist will ask: Who says? Why am I being given this piece of information and not another? What does that say about the teller of the story? How did s/he acquire the power to have their version of events believed by the general public? What sets of interests do they represent? Could that person have said that and got away with it at a different time in a different place? The questions are endlessly fascinating and frustrating.

C. Wright Mills published a short, seminal book in 1959 called *The Sociological Imagination*. In it he gave us the much-quoted idea that the basis of sociology is the analysis of the links between individual biography and historical change, between many disparate aspects of social life. Mills advises us to analyse the link between 'the personal troubles of milieu and the public issues of social structure'.[3] This deceptively simple formulation is actually very profound.

There is often major disagreement between sociological approaches, and what the key forces and interests are that drive social change. Contemporary sociologists tend to follow the work of one of the major social theorists. The main theorist referred to here, Pierre Bourdieu (sadly, recently deceased), was in fact deeply influenced by each of the three main 'founding fathers', Émile Durkheim, Karl Marx and Max Weber. Aspects of Bourdieu's approach will be raised throughout this book.

One's theoretical approach is connected to one's choice of research methods,[4] which can be primarily quantitative or qualitative. The former uses mostly statistical sources and analysis thereof, and the latter relies upon such techniques as in-depth interviews and observation studies.

There is a qualitative bent to this work, often backed up by quantitative reference to official statistics. I rely most heavily upon detailed interview data, as I am very interested in people's experiences of their society, and how they interpret it for themselves. A key concern here is to shine a light upon the meaning that people attach to aspects of their lives. Everything from the way we dress to the way we speak to the way we vote are socially constructed, a product of our perception of our place in society. As well as our individual characteristics, we operate in a shared socio-cultural world with those around us. This way of looking at things is termed a *constructionist approach*. This means that the main focus of my analysis is the negotiation

5

process involved in pitting one 'story-line'[5] against another in public discourse. Some story-lines become common sense and others virtually unthinkable. This suggests that social life is composed of differing claims emerging from different groups about what constitutes reality.[6]

According to this approach there is no ultimate truth because every social actor who contributes to debates on social issues ultimately is making specific claims for particular reasons. Social life becomes one big negotiation process when studied in this way. Power struggles arise because of these competing sets of claims about reality. This analytical approach is allied to a concern about structural power relations, because it is recognized that not all inputs are equal in their social impact and some inevitably become more influential than others.

Power has to be seen not simply in economic terms, but also in ideological terms. So while structures of economic power are seen to have a very definite social impact, culture and ideology also need to be investigated. This provides the tools needed to investigate if and how farmers, for example, could be viewed as subjects of power as well as objects thereof. This in turn helps us to view farmers as active agents within the social structures of rural life. The crucial concern here, then, is to deconstruct structures of power in a way that does not presume that they are uniformly imposed upon a passive population. This in turn allows us to examine precisely how EU policy informs farmers' lives, the actual social and environmental effects it produces, and how a certain set of ideas or knowledge about land use became dominant over others.

Some sociologists adhere more to the analysis of social structures and the constraining influences on people's lives, while others emphasize the agency that people can and do exercise in everyday life (the capacity to exert power over one's own fate). Agency is determined by the characteristics of both the people involved and the societies they live in: 'what individuals are doing only makes sense when we describe it in terms of the structures and institutions in which the activity is implicated'.[7]

Pierre Bourdieu negotiates this agency–structure dichotomy

by the application of his 'theory of practice'.[8] This suggests that looking at what people do in their everyday lives is the most meaningful way to find out how structures actually work. He argues that individuals are not separate from social structures, but that they are embodied or human representations of the complexities of structures. The ways in which people perceive their constantly changing positions in those social structures is determined by their *habitus*, which will be explained and discussed in more detail in chapter two. He says that 'in reality, agents are both classified and classifiers'.[9] He neither denies that observable, objective elements of social structure exist, like class, gender, ethnicity, and institutionalized power, nor that social actors can and do affect, create and recreate these seemingly impenetrable factors in subjective, personalized ways. He seeks to understand how people's common sense, their way of living their everyday lives, intersects with the operation of institutional and structural power. He is thus

> centrally concerned with how the various practical projects of different people, the struggles in which they engage, and the relations of power which push and pull them nonetheless reproduce the field of relations of which they are a part.[10]

Taking all of this complexity in account, it emerges that Bourdieu takes quite a pessimistic view – that social reproduction and stasis is far more likely than social change.

Intimately connected with this type of analysis is possibly the most major theme in contemporary sociology: the continuing and relentless process of globalization. We must acknowledge that the ways that people construct their own identities in everyday life are now governed by many disparate connections with groups, movements or fads that operate on a world stage. The stories that are told these days, by those in power and ordinary citizens alike, are often much more complex than they might have been in the past. We hear the term globalization a lot, and it is usually portrayed as either a good or bad thing, depending on who is speaking. Social scientists have been trying to grapple with the radical social changes brought about by globalization for some time now. We can witness global

---

Box 1: Globalization

Among the many definitions of globalization, one of the most satisfactory is that provided by Giddens:

the intensification of worldwide social relations which link distant localities in such a way that local happenings are shaped by events occurring many miles away.[12]

Globalization has social, cultural, economic, political and environmental aspects. Despite the fact that it began hundreds of years ago, there is a general agreement that it has intensified since the mid-twentieth century. There are more numerous and more intense flows of people, policies, images and ideas than there were in the past. This has been accelerated by the revolution in telecommunications and media technology. We now need to take this into account by extending our sociological analyses beyond national boundaries. Globalization exposes some groups, like indigenous peoples, to major risk; for others, like global entrepreneurs, it represents nothing but opportunity.

---

dynamics having profound local effects, in other words, the process of globalization having taken hold.[11]

Most analysts acknowledge that the principal master-process that governs globalization is that of capitalism. This is the prevailing economic system where property is privately owned, profit is the driving motivation and it is facilitated by the work of waged labourers. The current complex phase of capitalism has been explained in various terms. It has been described as the New International Division of Labour,[13] late capitalism,[14] disorganized capitalism,[15] flexible accumulation[16] and globalization.[17] Whichever term one prefers, some agreed characteristics of this phase are the following: the fragmentation of the labour process according to the resource needs and political dictates of transnational corporations; the increasing mobility of capital and its penetration into areas that were not previously industrialized; the questioning of the modernization mission; the revaluing of the idea and reality of 'local' areas;

the celebration of cultural diversity; increasing interaction between social actors in diverse locales; the emergence of many vigorous new social movements; more people becoming enmeshed in consumerism and a growing sense of global insecurity. While many of these aspects might appear contradictory, they all form part of the jigsaw that comprises contemporary social life, in which global dynamics can produce profound local social effects. Commentators argue as to the relative analytical importance of three forms of globalization: economic, cultural and political. It is very difficult to separate them, as each influences the other.

Economic globalization is visible both in the realms of production and consumption. TNCs fragment their production processes, so that they can derive the maximum benefit from availability of resources (like infrastructure and cheap labour) in different places. Another invaluable resource for them is the endless generosity shown by some countries, like Ireland. This generosity has taken the form of building Shannon Free Trade Zone, grants, tax relief, advance factories, guarantees of ununionized labour and relaxed environmental laws. States can choose whether to offer these incentives, hence the spread of TNCs is very uneven. Some favoured locations are South-East Asia, Mexico and Central America, Brazil and Ireland.[18] The African continent has never received more than 1.5 per cent of foreign direct investment in the past twenty years.[19] A simple model of economic globalization is therefore inaccurate, as states and groups of states in the form of trading blocs like the EU and North American Free Trade Agreement (NAFTA), can and do interact and negotiate with TNCs. Of course, this whole system is underpinned by the unseen workings of international finance, where frequently the only commodity traded is money itself. This system has often been exposed as being extremely fragile and skittish, in which one event like 9/11, an attack on the heart of western capitalism, can throw financial markets into complete turmoil.

Consumerism is the ideology that creates the demand for the products of capitalism. Globalized consumption can be seen in the increased availability of fewer established products. The power of the logo is standardizing tastes everywhere, in everything from fast food to soft drinks to jeans to cars. In Naomi Klein's path-breaking book *No Logo*,[20] she says that advertising executives see themselves as 'meaning brokers' who now create an association between a particular brand like Starbucks or Levi's with a whole lifestyle or attitude. The creation of the global consumer of such goods from Medellín to Mumbai is possibly the most significant cultural product of the globalization process. This new imperialism has been valiantly resisted by some groups, like the Fair Trade movement and the various strands of the anti-globalization movement.

Politically, globalization sometimes creates major challenges to the power of the nation state. Apart from the TNCs already referred to, many supra-national institutions exist that regulate political activity inside national borders. These range from the World Trade Organization (WTO) and the International Monetary Fund (IMF) concerned with trade and finance, to the North Atlantic Treaty Organization (NATO) and the United Nations (UN), dealing with security and peace-keeping. These international institutions have often been accused of disproportionately representing the interests of the richer countries: rather than reducing global inequalities, they often appear to deepen them.[21] Across the political spectrum, some globalized social movements have emerged in recent years that protest against the injustice of these inequalities.[22] These are movements that advocate the rights of women, ethnic minorities and indigenous peoples, or that aim to conserve the environment or address the problems caused by capitalist economics.

One of the core questions that plagues the globalization debate is how to conceptualize the interaction between global and local levels. Kalb argues:

> Empirical globalization outcomes … depend on social power relationships, local development paths, territorially engraved social institutions, and the nature of and possible action within social networks, and cannot be simply derived from any general framework.[23]

This analysis helps us to see the incredible complexity of contemporary social life, where simple systemic explanations for complex phenomena no longer suffice.

The rural regions of Ireland are now much more complex and differentiated as a result of globalization. The countryside is now subject to many competing constructions, as the dominance of farming interests can no longer be presumed. As it is used more for recreational and tourism purposes, new groups have emerged that represent different sets of claims upon its uses. It is not at all unusual any more to experience an area as both local (with its connotations of home, rootedness and emotional connectedness) and cosmopolitan (satisfying the need for exposure to wider economic and cultural influences) simultaneously.

A potential effect of globalization is that the more pressure that is exerted by economic and cultural forces, the more people may yearn for their local identity and seek to stress their difference to the homogeneity imposed from outside. People may need a human scale to complement the global, a means of inserting their own experiences, feelings and opinions into an often alienating world. This re-engagement with elements of local culture can be framed in an open or closed way. The local is appreciated by the person with cosmopolitan values as equal to other local cultures, and others are welcomed into their midst. Alternatively, the local is viewed by the xenophobe as something that needs to be protected and closed off. Hence outsiders with a different language or ethnicity are seen as a threat to the maintenance of a wholly mythical 'pure' local culture. These two positions are the two ends of a spectrum, and most people's views lie somewhere in between, depending on the particular issue with which they are faced. They could be open and tolerant toward one group and exclusionary toward another, for example, open to German incomers and closed to Travellers. Massey argues for 'a progressive sense of place'[24] in which the particularities of localities are celebrated, but outsiders not viewed as a threat.

The complex interaction between the global and the local in rural Ireland is captured tellingly in Patrick Kavanagh's famous poem 'Epic'.[25] The juxtaposition of the name of a large German city that was at the epicentre of global conflict with those of tiny townlands in rural Co. Monaghan is a bold statement of the weighty significance of local matters in the lives of country people like himself:

> That was the year of the Munich bother. Which
> Was more important? I inclined
> To lose my faith in Ballyrush and Gortin
> Till Homer's ghost came whispering to my mind.
> He said: I made the Iliad from such
> A local row. Gods make their own importance.

Antoinette Quinn argues that it is 'his finest and most subtle defence of old-fashioned rural subject matter'.[26] Seamus Heaney tells of his excitement at his first reading of Kavanagh's poetry. The details of country life, which he had 'always considered to be below or beyond books' were celebrated here, giving voice to local language and experience. In 'Spraying the Potatoes', 'the barrels of blue potato spray which had stood in my own childhood like holidays of pure colour in an otherwise grey field-life – there they were, standing their ground in print'.[27] This poetry has clearly resonated with a global audience – one hardly needs to hail from rural Ireland to appreciate it.

McMichael argues that it is an artificial exercise to separate the global and the local, 'as each template is a condition of the other'.[28] Social life is influenced by inputs from several co-existing levels, from global level to the EU, to national, regional, local and personal levels. It is not only a question of global forces percolating down but also of the extent to which local actors buy into cultural and economic globalization. The essential qualities of a particular place depend upon whose ideas about the present and future of that place become dominant at any particular time.

The *Mezzogiorno*, or south of Italy, has been analysed as a social construction:

> The South is much more than a geographical area. It is a metaphor which refers to an imaginary and mythical entity, associated with both hell and paradise: it is a place of the soul and an emblem of

the evil which occurs everywhere, but which in Italy has been embodied in just one part of the nation's territory, becoming one of the myths on which the nation has been built.[29]

Constructions such as these are necessarily multiple and complex because of the number of actors involved. Massey proposes that

> what gives a place its specificity is not some long internalised history but the fact that it is constructed out of a particular constellation of social relations, meeting and weaving together at a particular locus ... each 'place' can be seen as a particular, unique, point of intersection [of movements, relations, communications]. It is, indeed, a meeting place.[30]

If a place is defined by its inhabitants, then by definition no place can be understood in just one way. County boundaries in Ireland, for example, are in fact arbitrary divisions. They represent for Whelan:

> the successful imposition of the English shiring system initiated in the early medieval period and finally completed with the shiring of Wicklow in 1606 ... units of administrative convenience created out of a collection of Anglo-Norman lordships and Gaelic 'tuatha', 'nations' or 'countries'.[31]

Is it not odd that such an imperial product could retain such strong mental images and associations among Irish people in the late twentieth century?[32] These signifiers of identity are particularly important on two Sunday afternoons every September!

The Irish countryside is no longer simply a place where farmers work the land, but is now subject to several competing constructions. These constructions are not (solely) abstract metaphors, but have a very concrete impact upon the present and future of real places. Rural localities have often been seen as traditional and backward, compared to cities, which were seen as civilized and progressive locales. The urbanization of Irish society means that this construction of rural areas is now being overtaken by a vision of them as scenic, peaceful retreats from the hectic pace of urban life. More voices are also speaking up for the countryside, and an ecological consciousness is beginning to creep into official debates on development.

As cities like Dublin become more drawn into the dynamics of global capital, the resulting growth creates a place that is vibrant and exciting for those who can afford it, and a source of stress and expense for those who cannot. Rural living is seen by some as a potential solution to the uneven development pattern resulting from economic globalization. Agencies such as Rural Resettlement Ireland (RRI) argue that if more city people moved to the countryside, rural population decline could be reversed, alleviating the pressure for urban services at the same time. In the light of such problems as pollution, transport and waste management, rural localities now appear much more attractive than in the past, causing significant numbers of people to consider moving to the countryside. This is also leading to the increased urbanization of substantial towns like Castlebar, Clonmel and Tralee. This is concurrent with major pressures being placed on farmers by the liberalization of international trade on commodities like milk and beef.

The declining popularity of farmers and agricultural interest groups in Ireland and in Europe has presented a challenge to their political dominance. They are often viewed as 'cute hoors', clever at manipulating both Irish and EU politics to suit their own interests. This challenge has been led by representatives of the environmental lobby. The economic inefficiency, environmental destruction and social inequality associated with agricultural policy throughout the EU has led to a reformulation of the financial aid system for farmers. They are no longer viewed solely as producers of food, but also as environmental managers. The countryside is taking on a new cultural and symbolic significance, and the new meanings and valuations associated with rurality 'are continually contested as part of wider struggles over social power and resources'.[33]

Overview

Chapter one investigates the social impact of the Common Agricultural Policy (CAP) on Ireland and Europe. A new ideology was unleashed on farming people by the EEC and the Irish state after 1973, which encouraged them to be more 'modern' and more profit-oriented. This led to a growth in social inequality in rural Ireland, because the more a farmer produced, the more s/he was paid by the EEC. Those with the most assets have thrived on this system, while its competitive ethos has relegated many to the social margins.

Chapter two aims to demonstrate how useful sociological concepts can be in illuminating the coping strategies used by social groups and families in rural Ireland. The amount and type of power possessed by different people determines how creative they can be regarding their own futures. It is argued here that the entire family has to be taken into account when discussing farm survival, especially highlighting the vital role played by farm women. Their work in the home, on the farm and in the paid workplace is one of the crucial determining factors in creating a viable future.

Chapter three takes the reader through the processes involved in the reform of the CAP. This grossly exploitative and unequal system was clearly unsustainable, so some policy changes had to be made. As a result of these changes, a new strand has appeared within EU agricultural policy since the early nineties, which financially encourages some to farm their land in an ecological manner. This new arena, agri-environmental policy, appears to signal a major departure from the type of agricultural development that we have seen throughout the western world since the 1950s. During that era of modernization, issues such as water pollution and soil degradation were seen as unfortunate, but thoroughly excusable and inevitable, by-products of an industry that was vital to food security and the economy. The major agri-environmental scheme operating in Ireland is the Rural Environment Protection Scheme (REPS), which is outlined here. More recently, the complete decoupling of payments from production has been announced, and the ramifications of this will also be explored here.

Chapter four examines this current phase in farm policy in greater depth. It appraises the key ideas that are used to justify it, notably 'sustainable development'. This term, which is often taken for granted, is analysed at international and national levels. It is argued here that its main effect is to 'zone' different parts of the countryside for either productive or ecological functions, leaving the basic underlying structures untouched. The question of pollution control is also addressed here. It examines the difference that REPS makes in this regard, the practices of non-REPS farmers and the effects of the recent stricter implementation of the controversial Nitrates Directive.

Chapter five takes us outside the bounds of farming policy, and investigates some other claims that are made on rural space. The first of these is the contentious issue of rural housing. Some social groups and individuals are of the opinion that the regulations surrounding this are far too lax, and others think they are far too restrictive. These opinions are formed by one's orientation to the countryside; the current situation is analysed here from a sociological perspective. The other issue is organic farming, which has been very slow to take root in rural Ireland compared to some other European countries.

Broader landscape issues are addressed in chapter six, which deals with the debates surrounding conservation. The EU has decreed that the Irish state needs to set aside whole tracts of the countryside for ecological protection in order to mitigate the worst effects of commercial farming over the years. Landowners are not always happy about this, and this conflict is very evident in the current controversy about access for hillwalkers. The policy challenge here involves creating a code of practice that concurrently respects the concerns of landowners and also allows the increasingly urbanized population to get some enjoyment out of the landscape.

Finally, chapter seven constitutes a more detailed look at one region of the country that is close to many people's hearts – west Cork. Different groups of people make claims on an area like this, be they farmers, hoteliers, conservationists or second-home owners. In a small area like this, all of these groups live cheek by jowl. Each group constructs the region in ways that suit their own interests, and power struggles often arise because of these competing sets of claims. Two initiatives are highlighted, both of which engage with the process of globalization in very different ways. The first is Fuchsia Brands Ltd, an entrepreneurial branding initiative that attempts to confer a quality image and hence added value onto west Cork food produce and tourist services. The second is the west Cork branch of the Local Exchange Trading System (LETS), which is a barter movement whose aim is to delink as much as possible from what they see as a very harmful global economy.

Many of the quotes from farmers,[34] farm advisors, civil servants and politicians in this book were gathered from long interviews conducted for my doctoral research, in which the primary focus was the implementation of REPS. In the course of this work, however, opinions on lots of other issues were also gleaned: from farm incomes to planning issues to projections for the future. Farmers who inherited the land from their forebears (the 'inheritors') and those who bought their land relatively recently, actively choosing the rural lifestyle (the 'buy-ins') were interviewed. Many other sources are also used, from official statistics to newspaper articles.

The aim of *Land Matters* is to sensitize the reader to the complexities of social life in contemporary rural Ireland, and to break down barriers between people. The vast majority of us, despite our differences, will agree on one thing. We all want to see the maintenance of a thriving farming population, working the land that we all love so much in a manner that nurtures its priceless character and maintains a balance between the wild and the sown. Globalization, however, has created a new and challenging context in which farmers conduct business. A sociological perspective can help us to understand the pressures they are under, and the means by which they can ensure their survival. It explores the globalized flow of people, money and ideas. The most important analytical level we need to look at is the supra-national level, that of the EU, from which farmers receive most of their economic benefits, and where important political decisions are made. Focusing upon the CAP is the first step in attempting to divine the future for various social actors in rural Ireland.

chapter 1

The CAP and Social Inequality

A central concern of this book is how the lives of people living in local areas are affected by developments on the global political and economic stage. The EU is now one of the largest trading blocs in the world and its operations have a major impact upon its member states at every level. Joining the EEC in 1973 was the single biggest development to affect the lives of Irish farmers. Its Common Agricultural Policy (CAP) influenced what they produced, how they produced it, how much they produced and how much they earned for that produce. Peasant farming was to be discouraged and farmers were encouraged to be 'progressive' in their thinking. The rewards for those who were in a position to make this mental shift were enormous. The CAP completely changed the rules of farming and produced a whole new social profile in the farming population. This chapter takes a detailed look at the origins of the CAP and how it created the hugely unequal picture we see in contemporary rural Ireland.

From Agriculture to Agribusiness
For many years now, farming has become more integrated with the market and hence judged by rational economic criteria.

This process transformed farming from a way of life to a business, or from 'agriculture' to 'agribusiness'.[1] A central concept in the discussion of the modernization of agriculture is that of productivism (see Box 2). This has been shown to be the underlying ideology that has bolstered commercialized agriculture in the EU since the establishment of the CAP.[2]

Box 2: Productivism

Productivism advocates the relentlessly intensive use of land and natural resources, regardless of negative social or environmental effects. It is heavily influenced by the agriscientific establishment, whose aim is to constantly reach new frontiers in order to conquer nature. It is based on

the notion that a desirable agriculture is one in which there is progressive, self-reinforcing improvement in its total factor productivity through agricultural research and organisational upgrading of farms and agro-food firms.[3]

The effects of productivism are as follows:

- expensive, capital-intensive technologies progressively replacing human labour and the consequent increase in production and productivity
- farming activities becoming industrialized in terms of both inputs and outputs – meaning the fertilizers and pesticides put into the land – and the means by which goods are sold on the market
- production becoming concentrated in fewer and larger farm units, and also becoming more specialized[4]

Two further aspects of productivism can be added to this list. Firstly, commercial farmers are facilitated by banks and lending institutions because they are provided with favourable terms on financial credit. This leads to huge debt burdens being accumulated by farmers. Secondly, Irish farmers are becoming increasingly connected to large commercial co-operatives, especially in the dairy sector. The term for this process is vertical integration. Both of these latter factors are key elements that transform agriculture into agribusiness. They present farmers with economic inducements to increase their incomes at any cost, and also to surrender their beloved independence to large profit-making institutions. It will subsequently be argued here that EU agricultural policy has been a strong advocate of productivism, and that this is what caused the crisis of the CAP in the mid-eighties examined in chapter three.

This productivist ideology has been fundamentally guided by the idea of progress. This implies advancement, moving forward and leaving the past behind, making way for the new. Each new stage brings with it elements that are new and an inherent improvement on what went before. The ascription of the broad characteristics of traditional or modern to certain social groups and their activities has been central to post-WWII modernization, which depends upon this idea of progress. According to this perspective, the modern were worth investing in, while the others, the traditionalists, were a problem to be solved, either by changing or getting rid of them.

This division was made according to US ideological biases and has permeated development policy worldwide ever since. This philosophy was represented by the diffusion-of-innovations approach, which was very evident even in rural sociology in the sixties and seventies. The social psychology of individual decision-making was the most common subject of US research within this genre, whereas in western European studies of agriculture there was an emphasis on looking at means of changing the traditional cultural hindrances to the uptake of modern practices.[5] With the tremendous confidence that characterized the modern era, it was assumed that all vestiges of traditional life, these having to do with, for example, kinship relations, religion, and other local affinities, would inevitably be rendered obsolete on transformation to a modern condition. This approach is highly ideological because the underlying aim is to eradicate tradition in order to make way for the adoption of modern methods, and of course, the consumption of modern products such as fertilizers and pesticides.

However, it is argued here that it is more realistic to suggest that no social group, or indeed place, is wholly one or the other, traditional or modern, as those characteristics appear

simultaneously in all societies, presenting themselves in many wonderfully varied combinations. People living in rural areas are 'not outside historic time but a response to quite specific sets of historical social structures and historically conditioned environmental processes which are part of the specific experience of people living in a particular social and physical space'.[6] Rural producers may combine the most rudimentary farming methods that have not changed for centuries with the consumption of the latest biotechnology with a very postmodern seasonal migration pattern.

The Origins of the CAP

At this point, let us look at whether EU agricultural policy can be viewed as productivist. Article 38 of the Treaty of Rome stated that there should be a common policy for agriculture because:

> The maintenance of different agricultural systems would have led to distortions of competition which would have impeded trade and produced differences in the cost of food, and hence in the cost of living and in wage costs, which would have been prejudicial to true economic integration.[7]

At this time (the late fifties and early sixties) there was a large agricultural labour force in Europe, and many of those who worked the land experienced very low standards of living and had recent memories of food shortages and near-starvation. In the cases of Spain and Italy, this was exacerbated by the continued survival of pre-capitalist *latifundia*, reliant upon bonded labour and often run by absentee landlords. In other countries, such as France and West Germany, the dominance of industrial and urban-oriented development policies necessitated the introduction of a more equitable agricultural policy.[8] Article 39 outlined the aims of the CAP as follows:

1. To increase agricultural productivity by promoting technical progress and the best use of labour
2. To ensure a fair standard of living for the farming community
3. To stabilize markets

4. To guarantee regular food supplies
5. To ensure reasonable prices for consumers

It was stressed at the time that this should not threaten family farms or lead to overproduction. In fairness, it is perhaps only in retrospect that we can see the naivety of the signatories of the Treaty and that the five aims were actually conflicting. The CAP 'became a series of political compromises reflecting the different capacities of organised interests to influence the policy agenda'.[9] It was also probably the case that national interests took precedence over the desire for an effective union between member states, leading to the dominance of some countries, such as France, over others, such as Italy. Germany also managed to secure extremely high grain prices for its own farmers.[10]

It was a protectionist system that ensured farmers would always obtain a reasonable price for their produce by selling into intervention stock, and that would provide a protective buffer against the unreliability and volatility of international markets. This same protectionist approach was advocated by de Valera after gaining independence, but the national scale was too small to support it. At EEC level, however, it was deemed possible, for a while. Farmers received inordinately high prices for their goods, in fact. In 1987, EEC wheat farmers got a 51 per cent higher price than the world average, while butter prices were 80 per cent higher and beef prices 90 per cent higher. At that time, sugar-beet farmers were paid five times the going world rate for their sugar. This obviously hit Third World sugar producers hard.[11]

Various European Commission documents tell us that in 1962, when the CAP was established, the EEC was producing only 80 per cent of its food needs, so food security for Europe was a priority. This concern led to the encouragement of highly intensive farming methods in the member states, according to the dominant US development model of the time. The CAP, which has been the main influence on the Irish state since the early seventies, can be read as a modernization manifesto. The language encountered in EEC documentation from the sixties and seventies contains many references to stages of development, economic take-off and economic modernization. This productivist approach became the norm at this time, and as a result of this, agricultural production increased by 2.5 per cent p.a. between 1960 and 1980. The CAP therefore presented an archetypal example of a productivist development policy, with its inherent aims of expansion and intensification. This was achieved in the first twenty-five years of the CAP, with an average annual increase in production of 1.9 per cent per annum. The mechanisms used to achieve this were high internal agricultural prices, tariff protection at external borders and preference for European produce. Bonanno argues that:

> the CAP has made possible and convenient for many years a type of development in which increases in productivity were generated through a more intensive utilization of factors of production, such as land and capital, and a more than proportional growth of production.[12]

An EU policy document boasts:

> Within one generation, the Community was producing two to three times as much with two to three times fewer people, and from a considerably smaller surface area devoted to agriculture. Farming has developed more in the last 50 years than it did in the previous 2000.[13]

The CAP and Ireland

The commercialization of agriculture was already underway in Ireland by the seventies, but was reinforced by accession to the EEC in 1973. During the sixties, small farmers' problems were recognized by the state, and they had been encouraged to stay on the land by combining farming with off-farm employment. For example, the Second Programme for Economic Expansion (1964) initiated such schemes as the Small Farm (Incentive Bonus) Scheme, which was aimed at developing the potential of small farmers. Also, incomes were to be augmented by direct supports for farmers who owned under 30 acres.[14] Prior to joining, the objectives of Irish state agricultural policy, as stated in the Third Programme for Economic and Social Development

(1969–72), were 'increasing efficiency in production, processing and marketing, improving structures and securing better conditions of access to external markets'.[15] Alongside this was the dispersalist industrial policy initiated after the Buchanan Report of 1968, which advocated bringing industrial investment to key sites throughout the country, thus creating new development nodes. Accession to the EEC meant that many of these support schemes were withdrawn, and there was a more direct focus on payments for production rather than structural change. Since this time, the Irish state has been a very clear advocate of productivist capitalist agriculture. This became evident much later in the negotiations on Agenda 2000, with the Irish delegation in Brussels arguing that dairying and cattle production were indispensable to the Irish economy. This is indicative of a resilient entrenched productivist mentality, which has been responsible for the marginalization of small farmers for decades. The political aspects of these debates will be explored in chapter two.

This productivist ethos that quickly became the norm within Irish agriculture has emanated primarily from the Department of Agriculture and Food (DAF) and Teagasc. The DAF is over a hundred years old, predating independence. They were responsible for improving fishing and farming and educating farmers in 'modern' farming methods. Since accession, their main job has been to negotiate for more support for farming, and to manage and administer the vast sums that have been secured. Economist Alan Matthews argues that the need for a separate department for agriculture has come to an end, that it 'carries the danger that farmers' interests are over-represented in decision-making to the detriment of the national interest'.[16] Closely affiliated with the DAF is Teagasc (meaning 'teaching' or 'instruction' in the Irish language), who were formed in 1988 from an amalgamation of An Foras Talúntais (AFT) and An Chómhairle Oiliúna Talmhaíochta (ACOT). AFT was established in 1958 with monies from American Marshall Aid, and its purpose was the undertaking of agricultural research. ACOT came into being in 1980, and its role was in education, advice and research. Their main aim has been to modernize Irish agriculture, transforming it from small-scale peasant farming to larger, more efficient units that could compete on the international market. They have tried to instil in Irish farmers a productivist rationale and a respect for and knowledge of scientific methods. The use of scientific knowledge has been very important in legitimizing their existence in the public eye and also in consolidating their symbolic power within Irish agriculture.[17] To employ a military metaphor, Teagasc staff have been the foot-soldiers of the modernization mission of the Irish state. H. Tovey deems Teagasc to be:

> the institution most directly charged by the state for many years with the task of eradicating all but economic rationality in Irish farmers' considerations of how to manage their farms.[18]

The traditional–progressive dichotomy is close to the hearts of Teagasc staff. One young advisor who was interviewed here raised the issue of small farmers' outlook. When asked what he thought the problem with this was, this was his reply:

> It's not unusual to see small fellows looking on things very bleakly, that its all doom and gloom and 'how could you make a living' and all that. It's an attitude thing. You could have a progressive person in the same circumstances who'd look at it totally different. A lot of the smaller fellows are more fatalistic, they're accepting what's there without questioning what they could be doing or achieving … You're not going to get on with a non-progressive attitude.

It is clear from this quote that Teagasc differentiates between traditional farmers who are slow to change and adopt innovation, and progressive farmers who are the vanguard of positive change, adopting advanced technology and maximizing their profits. When asked what *he* meant by progressive, he replied:

> Intensive! In terms of creating more income for themselves and securing their future.

According to him, it is their own fault if they do not succeed. It is extremely condescending (not to mention inaccurate) to locate the main problem in farmers' socio-psychological makeup rather than in the monumental juggernaut of globalized agriculture. This represents a 'blame the victim' approach to understanding poverty. Farmers who are slow to take up the suggestions of farm advisory bodies are 'ideologically marginalised'[19] and their way of life undermined in public discourse. The fact that Teagasc has begun to charge farmers substantial fees for their services in recent years has met with widespread criticism. For example, the President of the Irish Creamery Milk Suppliers Association (ICMSA) said in 1995, when this change first occurred, that Teagasc were 'virtually abandoning low output farms to concentrate on directing its highly subsidized services to better-off farms'.[20] It is clear from this statement that some farmers' representatives also suspect that Teagasc are biased towards the commercial productivist farmer.

EU agricultural policy is based on the premise that there is only one linear path to economic progress, and that people everywhere respond to the same profit maximization logic. EU policy-makers view the farmer as a *homo economicus* who is free to, and does, act in the most rational manner to maximize his or her gains. Places and people who do not meet their requirements are deemed to be backward or 'lagging behind in development'. Advocates of productivism and modernization do not recognize that farmers may have other interests at heart besides profit maximization, or may interact with their land and the overall environment in more culture-specific ways than those dictated by dominant economic thinking. This became the official policy of the EU with regard to agriculture, aided by a highly ideological set of language that categorized farmers as 'development' and 'non-development'. In this dominant development discourse, modernization has been purported to be the tide that will raise all boats, bringing all participants in line with capitalist standards of progress and well-being, managed and distributed through the mechanism of the market.

The Tide that Raised all Boats?
Concurrent with this productivist ethos, EEC farming policy was extremely lucrative for Ireland in the seventies. Aid money poured into the country, with 81 per cent of all funding received

from Europe going to agriculture. Even though this dropped to 45 per cent in 2003, it still contributed vast sums of money to Irish farmers. The seventies was the most prosperous decade in the history of Irish agriculture, with real farm incomes doubling in 1978 compared to 1970. Entry into the EEC had the following advantages for the agricultural sector in Ireland:

- guaranteed stable prices for cereals, beef and milk products by selling into intervention stock
- lessened Irish dependence on the UK market
- a rise in farm incomes and land prices due to EEC support prices.[21]

On the other hand, this policy created a distinct regional pattern throughout the Community, in which most of the benefits accrued to intensive farmers in the more prosperous core regions of the EC, directly contradicting the EC's regional development objectives. Problems were created at this point that had to be corrected later. Considering that only 25 per cent of EC farms accounted for over 80 per cent of its total production at this time, one can easily see how reliance upon price measures as the main policy thrust could exacerbate the existing immense social inequality within the EC. It caused 800,000 agricultural holdings, or 10 per cent of total EC farms, to disappear between 1970 and 1987.[22] Bonanno asserts:

> Designed to aid small producers, these policies [of the CAP] have disproportionately benefited large producers, the most productive regions, and the budgets and profits of agro-food conglomerates.[23]

Larger farmers living in wealthier farming areas of Ireland were the primary beneficiaries of European financial assistance throughout these years. This severe social imbalance both within Irish agriculture and throughout Europe had to be rectified by subsidizing smaller farmers and those living in peripheral areas with direct payments. This was but one of the reasons that prompted the reform of the CAP in the mid-eighties. The modernization of Irish agriculture has led to increased social polarization between the larger and smaller farmer in Ireland. Joining the EEC was a major watershed point within Irish agriculture, marking the beginning of a new socio-political era that was to produce both positive and negative effects for different groups of Irish farmers.

EU agricultural funding has been administered and distributed through the European Agriculture Guarantee and Guidance Fund (EAGGF). This is divided into guarantee funds, or price measures, and guidance funds, or structural measures. The price measures guaranteed support for certain commodities, especially cereals, beef and milk products, and ensured fair prices for these goods through selling into intervention. Of these, milk is by far the most profitable commodity, leading to the dominant economic and political status of dairy farmers in Irish agriculture. It is argued here that this type of capitalist development strategy tends to favour the larger producer because they can adapt more readily by increasing their scale of production, while the smaller farmer often does not have the resources for substantial investment.

The second type were guidance funds, which were to fund structural measures. About 95 per cent of funds went to price support. Structural measures were less popular, perhaps partly because they operated on the principle of co-financing, where the state had to provide half of the funding for them. The central factor that has led to the polarized situation in rural Ireland is the overt differentiation between 'development' and 'non-development' farmers by EU policy.[24] This was evident in its price measures, because price supports tend to benefit the larger producer, going as they do to those sectors that are most capital-intensive and require large acreages. This widens the gap in wealth between the larger and smaller farmer.

The differentiation was also evident in its non-price, or structural measures. The 'commercial' or 'development' farmers in the better-off eastern half of Ireland were allocated most of the monies for farm improvement, while the poorer west secured nearly all of the headage payments in the seventies and early eighties.[25] One study showed that the top 17 per cent of farmers, those with 100 acres plus, secured 40 per cent of price support spending for themselves.[26] Even in the west, the better-off farmers got more EEC aid. There was a level of discrimination in favour of the west, but most of the monies were direct financing instead of incentives to actually develop their farming. Between 1975 and 1983, Less Favoured Areas funding (headage payments) comprised one-third of the total structural funding measures, and these payments also went to larger farmers, encouraging higher stock numbers. Fifty four per cent of structural funding went to a broader aid scheme for the west, termed the 'western package'.[27] While it gave farmers in the west a short-term financial boost, its social effects were more negative than positive. The farmers in this region were not encouraged either to invest in their farms or to be educated into some alternative farming enterprise. It also had negative environmental effects because it was thought to be the single most important cause of the serious overgrazing of the hillsides in the west.

It is now clear that both price measures and structural measures disproportionately benefitted those farmers who already owned the most resources – put simply, the more you had, the more you got. In the west, nearly half of all farms were under 12 ha, but only received 20 per cent of all grants and subsidies, while the 5.5 per cent who had farms of over 40 ha secured 17 per cent. In the east, one-third of all farms were under 12 ha, but received only 3.4 per cent of all subsidies, and the 20 per cent of farmers who owned over 40 ha received half of all grants and subsidies.[28] Smallholders were definitely discriminated against in this period. Most non-price, or structural, measures were concentrated on the young, business-oriented commercial farmer. Aid from the Farm Modernisation Scheme (FMS) was given only to those farmers who acquired development status by aspiring to be commercial farmers. Simple income maintenance went to the rest. This creates a particular social and spatial pattern, with the gap between richer and poorer ever-widening. Between 1955 and 1983, the incomes of farmers with 2–6 ha increased seven-fold, while the incomes of those with over 80 ha increased sixteen-fold.[29] In November 1998, the EU Court of Auditor's annual report found that 40 per cent of direct aid for arable crops went to just 4 per cent of farmers, 70 per cent to 10 per cent of farmers, and less than a third of all monies went to 90 per cent of farmers. It led them to say:

The indiscriminate granting of subsidies to agricultural operations which would be fairly profitable without subsidy cannot be justified.[30]

The gross inequality caused by this system is a dirty secret, which has never been exposed or addressed by the Irish state. Each successive government has simply focused upon getting as much as possible from Europe during its own reign of office and bowed to the pressure placed upon it by the farming organizations. The most recent estimate is that 60 per cent of total support goes to 20 per cent of farmers. About 146 farmers get over €127,000 each year in direct payments. Having revealed this fact, Matthews comments:

It is hard to believe that this money is needed to prevent them from falling into poverty, or that their contribution to rural society is so much greater than the average farmer's.[31]

This infamous social polarization is exacerbated by the decline of social services such as transport, health and education, which one can see in many rural areas.[32]

More and more Irish farmers are becoming integrated into the world of agribusiness, which requires fewer and larger farming units in order to be cost-effective. Agribusiness brings with it a whole set of ideologies about progress and proper land use, and a new expert discourse is becoming evident in the world of Irish farming. The more biochemical and technological products a farmer uses, the more modern s/he is perceived to be. This creates a 'technological treadmill', where the more one uses, the more one has to use in the future. Scott argues that in the EU, this treadmill 'revolve[s] as a result of the institution of guaranteed prices and an open-ended artificial marketplace'.[33] According to the logic of this treadmill, it is only when farmers adopt an industrial, commercial methodology that they are deemed to have caught up with their inherently more advanced urban peers. Intensive and highly specialized farming systems therefore need constant reinvestment to maintain their competitive advantage. The encouragement of this type of farming also clearly led to the crisis of overproduction that became a matter of urgency in the mid-eighties. The specialization and concentration of agricultural production has created a very clear spatial pattern throughout Ireland. Tillage is concentrated in Leinster in the east, and there has been a decline in cattle and dairying in the midlands, west and north-west. There is increasing divergence, then, between for example, the wealthy rural areas in Leinster and the south-east and the poverty of rural areas of the west of Ireland. This social and spatial polarization led Hugh Frazer of the Combat Poverty Agency to warn:

> Either we invest in trying to break this cycle [of poverty] or we resign ourselves to the gap between the advantaged and the disadvantaged widening.[34]

Farm Survival

It has now become clear that most of the benefits accrue to the top tier and scant attention is paid to the bottom tier.[35] It is commonly recognized that price supports have gone primarily to high-volume producers of milk in the prosperous areas of the country, so the CAP aim of fair income distribution was directly contradicted by this trend. Structural problems have not been addressed, because the policy on small farmers was almost solely dependant upon the welfare measure of headage payments.[36] The level of concentration of wealth that now exists makes it very difficult for smaller farmers to compete at all, and even to survive is an achievement. Many do not. In 1960 there were 278,500 farms in Ireland. By 1991 this number had reduced to 170,600. In thirty years Ireland lost 107,900 farm households. In 2004 the number of farms was under 120,000. This means that since 1960, about 3600 farmers left the land each year. This shocking trend is likely to continue. It has been projected by a high-profile task force in the agri-food sector that only 25,000 commercial farmers will remain in 2015, including 15,000 dairy farmers, 5000 in drystock, 2000 arable producers and 500 in pigs. There would be another 35,000 part-time farmers, reliant on off-farm income.[37] At current rates of departure from the land, this scenario is easily imaginable. The social and environmental effects of this trend are worrying indeed.

Farm size in Ireland is still relatively small. There are still only 6700 farms in Ireland of over 80 ha. Table 1 compares the absolute number and proportion of farm holdings by size in Ireland, in 1855 and 1991, and the percentage change within each category over time.

Table 1: Number of Holdings by Size, Ireland, 1855–1991

Acres	1855	1991	1855 (%)	1991 (%)	Change (%)
1–5	61,800	4000	14.7	2.3	-93.5
5–15	127,200	18,000	30.3	10.6	-85.8
15–30	101,500	31,000	24.2	18.2	-69.5
30–50	55,600	38,500	13.2	22.6	-30.7
50–100	44,900	48,700	10.7	28.6	+8.5
100–200	19,300	23,200	4.6	13.6	+20.2
200+	9200	6700	2.2	3.9	-27.2
Total	419,500	170,100	100	100	-59.5

Source: Derived from CSO, 1997. Note: Historical data uses acres as a land unit. 1 ha=2.5 acres.

Overall, the number of farms dropped by 59.5 per cent, from 419,500 to 170,100 between 1855 and 1991. There was an increase in the number of farms between 30 and 100 acres from 23.9 per cent of all farms in 1855 to 51.2 per cent in 1991. In 1855 only 6.8 percent of farms were of 100+ acres (40+ ha). By 1991 this has risen to 17.5 percent. In the 200+ acres (80+ ha) category, we witness little or no change, with only a very small rise from 2.2 per cent to 3.9 per cent.

Commins and Keane (1994), from their analysis of the Teagasc National Farm Survey, estimate that out of a total of 159,000 farms in existence at that time only 50,000 were economically viable, or about one-third of all farms. Among those 109,000 that were not, the presence or absence of another income on the farm was the main criterion for survival. Eighty thousand of these farm households had no other income and of these, those 32,000 who were 'demographically non-viable' had an average annual farm income of £2400 (€3047) in 1993.[38] Those 48,000 who were 'demographically viable' had an average annual farm income of £4600 (€5840). The vital role of farm women is a major factor here, reproducing the farm both demographically and economically. In 2001, just seven years later, Connolly found that 120,300 farmers remained in the country.[39] There were 41,300 viable full-time farmers left, 36,400 non-viable part-time farmers, and 42,600 non-viable transitional farmers. The average income on full-time farms was €22,900, compared to the average industrial wage of €24,500. Between 1994 and 2001, farm numbers declined by almost 40,000, from 159,000 to 120,300. The number of farms of over 50 ha remained static and those of under 20 ha showed substantial decline.[40]

Such statistics can hardly be explained away as a natural retreat from the land, but are instead best understood as a direct result of market penetration of the countryside. Social opportunities tend to be distributed more unevenly by the market, because if more food is produced than is required, there is a continuous downward pressure on prices over time, creating more competition between farmers to maximize their gains and to be more uncompromising in their methods.[41] This pattern of social polarization would appear to be replicated throughout the EU, with increasing divergence between rich and poor areas within states, and between rich and poor states. In a detailed survey conducted by the Arkleton Centre in Aberdeen, they found that the CAP tended to benefit richer regions with lower unemployment rates, and areas of highest population growth. This led the authors to conclude that the agro-industrial model still dominates the CAP, and the thinking of those who administer it.[42]

There is a large number of very small farms throughout Europe, and especially in the countries of the southern periphery. Nearly all farms in Italy, Portugal and Greece are under 20 ha.[43] The fact that the vast majority of European farms are this small leads one to suggest that a farming policy that persistently encourages large-scale commercial farming through EU pricing mechanisms appears to be at least misguided, if not completely unjust. The fate of the European peasant farmer, and indeed the places in which they live, looks bleak indeed when one looks exclusively at these statistics. Productivist policy has made it very difficult for them to compete, or even to survive. There is no doubt that the largest proportion of EU funding has gone to the biggest producers. It has been estimated, for example, that in the 1984–6 period, across the EC, larger farms received 15 times the support that smaller farms did.[44] Over 80 per cent of all sales have been produced by only 25 percent of all farms. These large farming units have disproportionately benefitted from EU financial aid to the agricultural sector.[45] This general pattern has led Symes to assert that:

> a profoundly discriminatory and polarised structure of production has emerged with highly favoured, large farms close to the economic centre of the Community and a myriad of small farms struggling for survival in the southern and western peripheries.[46]

The farmer who did not meet the EU criteria of resource base, demographic profile or economic aspirations was excluded, both ideologically and financially, by the mechanisms of both price and non-price measures. The power map in Irish agriculture, as apparently throughout Europe, is as a result a picture of the dominance of the technologically oriented farmer and the marginalization of those who cannot compete in this undemocratic environment. The ideology of modernization is the root cause of this profoundly unjust form of social differentiation. C. Eipper, in describing the social changes that resulted from this in rural Ireland, asserts that:

> As the ideology of progress, equated with commercial competence and modernization, permeated the consciousness of the young, the authority of age and tradition was forced to adapt itself to the inexorable advance of the authority of the market.[47]

Dairy Farming

These trends are especially visible in the dairy sector. As far back as 1982, Tovey hinted at the importance of the study of the changing farming culture in dairying. She questioned why small milk producers were marginalized by a system that was designed to help them. She argued that too much research had absorbed the idea that the characteristics of small farmers must be changed in order to catch up with their bigger neighbours, that the problem was with the individual rather than structure. The way dairy farming had become so reliant on capital-intensive scientific methods served to marginalize the small producer. They become dependent on advisors and on loans from banks, and on a different, more scientific way of thinking, exemplified by ACOT (now Teagasc).

Since then this trend has escalated and the number of dairy farmers in the country is now reduced to just over 25,000. Before quotas were introduced in 1984, there were 86,300 dairy farmers in the country. There has been an enormous drop in the number of small- and medium-sized milk producers: it is now at 29 per cent of the 1983 figure. In 2002 alone 1400 dairy farmers went out of business, which was 5 per cent of the overall total at the time. The average amount of milk quota allocated to each supplier is rising every year, for example, from 170,700 litres in 2000 to 201,500 litres in 2003. There was an increase of 7 per cent in quota size in just one year, between 2002 and 2003. Five per cent of the country's dairy farmers (1204 in number) now supply 15 per cent of the country's milk.[48] The dairy industry is crucially important, accounting for one-third of Irish agricultural output. It has been envisaged by agronomists that by 2010 there will be only 20,000 dairy farmers left, producing an average of 320,000 litres each.[49] It is assumed that milk quotas will be abolished in 2008, creating a totally free market situation. In that context, there will be a constant challenge to satisfy market demand for diversified milk-based products, such as milk powder and various types of cheese.

There now seems to be an inverse relationship between the number of dairy farmers and the amount of milk produced. While one can glibly write this off as just 'the way things are', the effect of these statistics on the social fabric of the countryside remains an untold story. The agronomists and economists have long dominated debates on farming, with the sterling work produced by rural sociologists being largely ignored. There is also, of course, an inherent contradiction between the ongoing trend of intensification of production and the EU environmental controls that will inevitably become stricter in the near future. Environmentalists say that on a global scale, there are three major problems associated with industrial agriculture: an escalation in the price of oil and gas, the very real threat of disease in crops, and scarcity of water.[50] Advocates of productivist agriculture, epitomized by the large milk producers, are adopting an arrogant head-in-the-sand approach to this most momentous of issues. Any other consideration is sacrificed to the maintenance of their top position on the league table of farm incomes.

Farm Incomes

This issue is a difficult one, with farmers constantly claiming that their incomes are low relative to the amount of work they put into their land. This is countered by the amount of money given to them in direct payments by the EU. To the casual observer, there is evidently an underlying contradiction somewhere in relation to farm incomes. The main available source that reviews aspects of Irish agriculture is Teagasc's annual National Farm Survey. The 2003 version[51] gives an average Family Farm Income (FFI) of €14,925, which is down 5.8 per cent on 2001. However, such an average figure reveals little, as it subsumes a whole variety of sizes and types of farm enterprise. It also excludes off-farm income, so it probably does not equal total household income. The survey's differentiation between sectors is more interesting, telling us that those in dairying earn €28,084 p.a., those in tillage €21,884, sheep €12,354, and cattle €7752 p.a. Dairying is still by far the most profitable category and the keeping of dry cattle is generally only for those who have another job or are too elderly to compete in any other type of enterprise. It also highlights the reliance of farmers on direct payments and off-farm income earned by the farmer him/herself and his/her spouse.

The survey also tells us that 40 per cent of all farms had an income of less than €6500. However, another economist says that farm households are not especially poor anymore, that they now make up less than 5 per cent of all households in the state who are classified as poor, and that the situation was likely to be improved by the welfare programme, the Farm Assist Scheme.[52]

These bald economic statistics obscure the reality of everyday life and the stresses farmers are under. Naturally, these figures exclude those who have already left farming in recent years and no longer define themselves as farmers in the census, but who still may be experiencing rural poverty. If small and medium farmers are not a high-risk category for poverty, then why are so many of them making the wrenching decision to seek other work? After all, it is not quite like any other career choice because of the huge emotional investment in the land that has often been in the family for generations. With so many leaving farming each year, this releases quite a few into the workplace who probably have few qualifications and low levels of education. There is little or no knowledge of or support for this silent and beleaguered minority who have little bargaining power with employers. With privacy being such a treasured value in rural Ireland, these people have become accustomed to quietly making the most of their meagre circumstances. Consequently, they are probably likely to make do with relatively poor pay for casual labour in their local areas. The types of work undertaken by half of all part-time farmers is in agriculture itself or the construction industry. This new reserve army of labour will be an important group to observe in years to come.

Conclusion

This chapter has detailed how the CAP created a two-tier rural Ireland, and indeed Europe. When the CAP was introduced in 1962, it was deemed necessary in order to achieve self-sufficiency in food production and to curb the severe social inequality that was evident both within and between the regions of Europe. Its main goal was to modernize European farming and introduce highly intensive farming methods. It has transpired now that it disproportionately rewarded aggressively productivist farmers at the expense of European taxpayers. This has created a culture of greed, which has led to the gradual demise of small farmers who are often ill-prepared for their off-farm futures. It is only those who can compete commercially or who have been educated to prepare them for some other type of work who can be optimistic about their long-term futures. Having discussed the economic effects of the CAP, let us now look to a more sociological approach to try to understand how some people can acquire the social power to do very well in rural Ireland, and how others are relegated to the ranks of the poor and the exploited.

chapter 2

Survival Strategies in Hard Times

Passing through the Irish countryside, it is quite common to see estate agents' signs advertising small farmsteads for sale. We saw in chapter one that the number of farmers in Ireland who can survive in these competitive times is indeed dwindling. However, when one considers how fast the pace of change has been in Irish agriculture, one might expect farmers to be in a much weaker economic position than they actually are. While farm decline is indeed significant, if we were to read economic signals alone, we would expect the situation to be much worse. Farmers have always demonstrated great tenaciousness and adaptability. This spirit of survival appears to be common throughout the western world. This chapter proposes an analysis of Irish farmers that is informed by the key concepts of habitus and capitals bequeathed to us by Pierre Bourdieu. It argues that this framework can help us to understand in a deeper way the means that different groups of farmers use to make the most of their circumstances. It begins by briefly reviewing the classical authors on the subject of the social impact of the capitalist penetration of agriculture. It goes on to insert Bourdieu's theory of practice into the contemporary debate, and finally, to apply this framework to the case of Irish farmers.

2. Survival Strategies in Hard Times

Agriculture and Capitalism

Contemporary debates on farming, and social life in general, are still informed by classical debates from the nineteenth century. This is because the questions raised then have a perennial interest and relevance. There was a major disagreement between Karl Marx and Alexander Chayanov, for example, on the social impact of the commercialization of agriculture. Marx, in the *Eighteenth Brumaire,* believed that the fate of the European peasant was sealed, that he would be squeezed so hard that he would ultimately disappear and join the ranks of the urban working class.[1] Marx lacked faith in European peasants because they were too individualistic and independent to develop common class interests, (in)famously likening them to how 'potatoes in a sack form a sack of potatoes'. In his terminology, they did constitute a class in itself, but not a class for itself. This means that they share objective conditions, but not consciousness. He continued: 'They cannot represent themselves, they must be represented.'[2] This denied the peasantry any agency or capacity for political organization.

Lenin took the Marxist analysis of the peasantry one step further, however, by focusing upon the processes that created social differentiation among the peasantry. Differences emerged between the rural entrepreneurs who hired wage-labour in order to expand their enterprise, and those who were landless and remained as rural workers, or proletarians.[3]

Conversely, Chayanov tried to explain the persistence of family farming despite the pressures imposed by capitalism. He disagreed with Marx that the peasantry could be understood just as petty commodity producers. He instead argued that they constituted a specific type of economy, which showed a fundamentally different motivation to that of capitalism: securing the needs of the family rather than increasing profit.[4] Peasants, and their family labour, were deemed external to capitalism, unaffected by the economic context. They therefore merited separate consideration. The peasant producer did not aim for expansion; firstly, because of the natural limits imposed by the land upon concentration of production, and secondly, because of the finite nature of their labour and resources. He presented his argument thus:

> the degree of self exploitation is determined by a peculiar equilibrium between family demand satisfaction and the drudgery of labour itself[5]

Therefore, considering how limited their labour power was, their efforts eventually reached a point beyond which it was no longer worth striving.

There is value in both the Marxian and Chayanovian perspectives, the former emphasizing the inexorable pressures placed upon small farmers by capitalism, the latter emphasizing the differences between peasants and other workers under capitalism. A more contemporary approach has helped to breathe new life into rural sociology by integrating peoples' own opinions and experiences into the analysis. This approach recognizes the historical and geographical diversity of peasant economic and cultural practices. It helps to illuminate the tactics that people use to ensure their survival. We can see this in the work of van der Ploeg, where he argues that farmers negotiate agricultural markets and technologies by utilizing their 'cultural repertoire', or local cultural constructs. He argues that farmers develop 'styles of farming', which are locally constructed responses to global markets.[6] This is a micro-analytical approach, which deems the agency that individuals exert to be the most vital area on which to focus. This is highly compatible with Bourdieu's theory of practice (see the introduction).

Habitus and Capitals

The aforementioned agency/structure question is a major concern of Bourdieu's. He combines studying social structures with analysing people's own opinions and experiences of those structures. Sociologists term these two approaches structural and phenomenological. He explains that 'in reality, agents are both classified and classifiers'.[7] He neither denies that observable, objective elements of social structure exist – like class, gender, ethnicity, and institutionalized power – nor that people affect, create and recreate these seemingly impenetrable factors in subjective, personalized ways. He seeks to understand how people's common sense intersects with the operation of institutional and structural power. He is thus:

> centrally concerned with how the various practical projects of different people, the struggles in which they engage, and the relations of power which push and pull them nonetheless reproduce the field of relations of which they are a part.[8]

His main contribution to this debate has been the extremely valuable concept of habitus.

Box 3: Habitus

Habitus is the unconscious means by which people place themselves, or auto-select a location, in social life. One line from Bourdieu's *Outline of a Theory of Practice* sums it up succinctly: 'the dialectic of the internalisation of externality and the externalisation of internality'.[9]

By this he means how people, in their everyday lives, recognize that they hold a certain position in their society in relation to others, and act accordingly. Structures are therefore internalized and are far from external to the individual, but are literally embodied in each person or group of persons.

His approach leads the sociologist to observe in detail what people actually do, focusing upon why they do certain things and not others. He stresses that this is the key to finding out how structures actually work in everyday life. The individual's experience of the social world serves to fence off certain possibilities, and people usually behave as they are expected to by society, inside rather than outside the social structures in which they find themselves. The habitus of a person, or group of persons, prescribes certain courses of thought and action, laying down the divide between the possible and the impossible, the appropriate and the inappropriate. Bourdieu asserts:

> Dominated agents ... tend to attribute to themselves what the distribution attributes to them, refusing what they are refused ('That's

not for the likes of us'), adjusting their expectations to their chances, defining themselves as the established order defines them … condemning themselves to what is in any case their lot.[10]

This can help explain the stamina of the inequality inherent within advanced capitalist societies. Habitus has been defined as 'the system of durable and transposable *dispositions* through which we perceive, judge and act in the world'.[11] It is how people unconsciously assess what are appropriate (and inappropriate) behaviours and aspirations for them. This leads to the conclusion that people are usually more likely to accept the status quo than to attempt to change it.

Bourdieu is sometimes accused of having a conservative bias, but this is because he asks the most probing questions about just why it is so difficult to achieve social change. He sees unequal social relations as a problem to be solved. He has a basic desire for positive social change, and thence asks why it does not happen as often as he would like. Why is it that we witness social reproduction more often than social change? He deems the most important question for sociologists to be 'the means by which systems of domination persist and reproduce themselves without conscious recognition by a society's members'.[12] Society reproduces itself with the least conflict when there is an almost perfect match between what is provided by society for an individual, and what they expect for themselves, when 'the agent's aspirations have the same limits as the objective conditions of which they are the product'.[13]

He stresses that society can be reproduced without strategic intention, that life can go on for most people in a state of 'conductorless orchestration'. The operation of power relations may not always be intentional, but nevertheless 'each agent, wittingly or unwittingly, willy nilly, is a producer and reproducer of objective meaning'.[14] Each person carries around an 'embodied sedimentation' of social structures, a habitus, whether they realize it or not. This does not prevent social or personal rebellion or expressions of dissent, especially considering the cultural effects of globalization, but it is nevertheless a shadow that falls to some extent on all of one's actions.

Sometimes this shadow is short, sometimes long, sometimes visible, sometimes invisible, but it is nevertheless omnipresent.

The concept of habitus therefore helps us to negotiate the fractured terrain of the agency/structure question. Bourdieu summarizes it like this:

> the habitus, the product of history, produces individual and collective practices, and hence history, in accordance with the schemes engendered by history.[15]

It is a notoriously difficult, slippery idea, which has been criticized by some for its lack of clarity. However, one person's 'lack of clarity' is another's 'adaptability' in dealing with 'imprecise, fuzzy, woolly reality'.[16]

Within Bourdieu's approach, the characteristics of one's habitus can be defined in terms of the possession of different types of capital. He proposes that there are other means of achieving social power besides economic wealth. It is in this area that Weber's legacy is most evident in Bourdieu's opus. Weber contributed the important concept of status to the analysis of social class. He was the first to suggest that there were means of increasing one's social power that had very little to do with economic assets, but had more to do with culturally based characteristics like everyday behaviour, reputation and social connections. If the possessors of a certain kind of symbolic wealth or status are part of the dominant group in society, how do they get to that social position, and more importantly, how do they stay there? If economic wealth or capital is not the only, obvious, means of achieving social power, what are the other forms of capital that can be used for this purpose?

Bourdieu defines the different forms of capital as 'those properties capable of conferring strength, power, and consequently profit on their holder'.[17] The key point is that the three main types of capital – economic, cultural and symbolic – can be converted into each other and cashed in on a market:

> Because culture is embodied (in language, action, style), objectified (in art works, buildings, books), and certified (in educational credentials), it can be exchanged, converted and reconverted to produce power and domination.[18]

It is even difficult to separate out the different capitals, as each is inextricably linked with the others; each serves to reinforce the others, in various ways and in various contexts.

While economic capital, the ownership of money, property, land and other financial assets, is still undoubtedly a vital source of power, it may not be enough in some cases to get things done the way one wants them or to exert influence over others. It may depend on how one has made one's money as to how much status it can buy. If one wins the Lottery, it probably will not buy one a place among the rich and powerful.

One may instead need vestiges of cultural capital to achieve this position, which is determined by the inheritance of educational and social resources, the main motors of social mobility. We can see in contemporary Western society that the cultural resources of education and prestige are becoming increasingly socially significant. One's lifestyle, values and attitudes are primarily shaped by cultural capital. One can accumulate cultural as well as economic capital through attending those schools and mixing with those people who are socially valued. Bourdieu says that academic qualifications are to cultural capital what money is to economic capital. He is interested in the production of symbolic goods, but he is even more interested in 'the system producing the producers', whose machinations are conducted in secret.[19] The power of academic qualifications means that educated people often just have to quote their degrees or titles to achieve social legitimacy in most circles. Qualifications and cultural capital are intimately connected, so the system of domination is legitimized in a self-perpetuating, circular manner.

Finally, symbolic capital refers to the domain of the prestige and renown that one, or one's family, possesses because of one's social connections and relationships with other power-holders. It shapes how systems of domination are recognized and accepted as legitimate by the less powerful in society:

> In the struggle to impose the legitimate view of the social world ... agents yield a power proportionate to their symbolic capital, i.e. to the recognition they receive from a group.[20]

2. Survival Strategies in Hard Times

Authority is conferred upon an individual or group when their control over the internal workings of the society is virtually unquestioned. Bourdieu found this to be a very important source of power among the Kabyle, the ethnic group in Algeria that was the subject of his initial anthropological research. Every opportunity is used by the Kabyle to demonstrate symbolic capital, especially on such occasions as weddings. Money is spent on items that appear irrational to the outsider, but in fact they are highly rational 'in a good-faith economy in which good repute is the best, if not the only, economic guarantee'.[21] The family name is at stake, and this is vitally important in a society where matrimony is one of the primary means of accumulating the interlinked assets of economic and symbolic capital.

Social actors can acquire social power through the potentially complex interplay of the main three types of capital – economic, cultural and symbolic – within specific spheres. Through their habitus, some actors are predisposed to possess and exercise power, while others are predisposed to recognize the status quo as legitimate and submit to it. No set of knowledge becomes dominant without an audience that recognizes it as such, through the internalization of structures. In this way, the oppressed often are complicit in their own oppression:

> Dominated individuals are not passive bodies to which symbolic power is applied, as it were, like a scalpel to a corpse. Rather, symbolic power requires, as a condition of its success, that those subjected to it believe in the legitimacy of power and the legitimacy of those who wield it.[22]

Some people will be intimidated by an action of a symbolically powerful actor and others will be oblivious to it or simply ignore it, because of their pre-inscribed habitus. The context is important too. A small-town businessman might be extremely powerful in his hometown, for example, but his combined resources may look relatively paltry when he attempts to infiltrate the horse-racing fraternity at Ascot. At this point, we turn to applying this conceptual apparatus to providing an alternative social profile of farmers in Ireland, leading to an analysis of their survival strategies within capitalist agriculture.

Irish Farmers: Habitus and Capitals

So far, we have seen that gross inequality exists among the Irish farming population. We need a sociological means of explaining why some are better able than others to survive the vagaries of capitalism. Bourdieu's theoretical framework helps us to do just that, to see that the agency exerted by farmers in Ireland and elsewhere is strongly circumscribed by their habitus and the amounts and types of social power, or capital, they possess. Habitus is the catalyst in the relationship between objective and subjective components of social group (or class) identity and we must know how it works before we can know what to expect from social actors. In his reflections on peasant culture, Ortiz asserts:

> Their perceptions become internalised and institutionalised and constitute the lens through which they view the real world, even when the real world changes and offers them more rewarding opportunities.[23]

Keeping this in mind, it is useful at this point to discuss the capitals possessed by farmers, which comprise their habitus and determine their chances of survival.

Economic Capital

As is generally the case in social science, the poor are studied more often than the rich in rural sociology. More is known about the lives of smaller than larger farmers. We can gain access to broad income statistics for both groups, but qualitative research, informed by observation and interview data, is more commonly conducted among smaller farmers. Wealthier people are less likely to be approached by researchers, and probably more likely to refuse to cooperate if they are approached. The data presented here is a result of this research context.

When discussing the economic assets of farmers, it is only natural to turn first to farm size. By European standards, this is relatively small in Ireland. Thirty-one per cent of all farmers owned under 12 ha in 1991, over half (51.5 per cent) of all farmers owned between 12 and 40 ha, so in total 82.5 per cent owned under 40 ha of land. This leaves 13.6 per cent owning

40–80 ha and just 3.9 per cent owning 80+ ha.[24] Referring back to Table 1, we saw that small farmers have dramatically declined in number over the years, whereas farmers who own over 40 ha have increased and stabilized their economic status. These large farmers also own more technology: e.g. in 1990 the value of machinery on farms over 100 ha was 7 times the national average.[25] It is, however, misguided to focus solely on farm size. Over the years, concentration of production has actually become a much more significant phenomenon than concentration of land ownership. In 1990, the top 20 per cent of farms were responsible for 39 per cent of agricultural land and 60 per cent of farm output. However, notwithstanding farm size, income earned largely depends on the type of farming enterprise, as can be seen from Table 2 below.

Table 2: Family Farm Income (FFI) (€) by Farm System and Farm Size

Size (ha)	<10	10–20	20–30	30–50	50–100	>100	Hill fms	All
Dairy	–	12,500	22,700	28,900	44,300	70,800	17,100	28,100
Dairy /other	–	–	4700	19,550	39,000	57,500	–	25,200
Cattle	2500	4500	6200	12,700	19,100		8300	7800
Cattle /other	5100	3300	7100	12,100	25,000		7200	9500
Sheep	–	3800	10,400	15,200	27,200		12,900	12,400
Tillage	–	–	8300	15,200	28,000	60,900	–	21,900
All	3500	4700	9700	18,000	32,900	54,400	11,700	14,900

Source: Teagasc National Farm Survey, 2002.

The top tier of farmers are well-paid for their (albeit considerable) efforts. The average FFI for dairy farmers (at €28,100), is double the average FFI for all farmers (at €14,900). Both within and between farm categories, there is considerable variation in FFI. For example, a farmer who owns 30–50 ha (which is the average farm size range in Ireland) can expect an FFI of €12,100 if s/he rears cattle, and €28,900 if s/he is in specialist dairying. A dairy farmer who owns 10–20 ha earns €12,500 p.a., whereas if s/he owns 100+ ha, s/he earns €70,800. The single category with the highest income by far is this latter group, dairy farmers with over 100 ha.

The Farm Survey also tells us that 8 per cent of farms had an FFI of over €40,000, and three-quarters of these were in dairying. Also, 25 per cent of farms achieved a gross margin of €1000/ha in 2002, and two-thirds of these were dairy farmers. FFI usually increases with farm size, as does income per ha. These large (especially dairy) farmers have accumulated a lot of machinery, staff, and business acumen over recent years. They have experienced very few limitations on their activities, which have been dedicated to capital accumulation on a substantial scale. Their habitus reinforces that they are valued and useful members of Irish society, and they will adopt whatever farming practices are necessary to increase their profits. They have the backing of the banks and of the state. Joining the EEC benefitted these large dairy farmers greatly, and continues to do so. The pricing mechanism dictates that milk is still by far the most profitable commodity. The risks associated with becoming so deeply entrenched in capitalist agriculture are clearly offset by some substantial benefits.

One must conclude here, however, with a caveat, that this group's power is increasingly circumscribed by the demands of the large retail sector and the agri-food industry. The low prices charged for milk by Lidl, and the resulting price wars with other supermarket chains, is already forcing farmers to sell milk below cost of production. The presence of these chains, the so-called 'hard discounters', is creating a very tough retail environment for farmers. The new EU environmental demands, especially relating to the Nitrates Directive, are also likely to present a major challenge to their activities in the very near future. This will be discussed in much more depth later.

The social status of smaller farmers is currently declining in the broader society because of the increasing integration of farming into the world of agribusiness. Smaller farmers, who constitute the numerical (if not the political) majority of Irish farmers, have been marginalized by the productivist policy initiatives of the Irish state since the fifties and especially since accession to the EEC in 1973. Only the commercial producers are deemed necessary to sustain the growth of the food sector, while the rest are either superfluous to requirements or reallocated the role of the producers of environmental and/or tourist goods. Irish society is also becoming increasingly urbanized, with urban commuter belts expanding beyond all expectations. More and more land will thus be zoned to satisfy the burgeoning housing needs of the Irish (sub)urban population.

Smaller farmers usually operate farms based on beef cattle, sheep, and suckler cows – the most vulnerable sectors with the lowest incomes. As we saw in Table 2, a sheep farmer with under 20 ha can expect an FFI of €3800. If they have a small dairy enterprise, they cannot increase their quota size, as they are stuck with the allocation they received in the mid-eighties. Their only option is to lease quota, and now compulsorily, the accompanying land. The prices charged for these are often prohibitive for this size of farm operator. The main policy initiative adopted by the state that aims to help this group is REPS, the agri-environmental scheme. Paying smaller farmers to farm in an environmentally friendly fashion is the only welfare approach that is now politically acceptable at EU level. The scheme is discussed in detail in chapter four.

Smaller farmers have had to use their habitus to adapt to this harsh reality in a number of ways. In economic terms, they have been very slow to let go of the land and sell up, even when they have found other sources of income like EU transfers or off-farm employment. This will be done when absolutely all other options have been exhausted. This has been attributed to the pride that comes from land-ownership in a relatively recent postcolonial society.[26] In the first three months of 2003, only 157 land transactions were recorded, and the average sized land parcel put on the market was *c.*10 ha.[27] This reticence is one of the main reasons that small farming has not abated to perhaps quite the extent that many commentators thought it would. They do their best to retain ultimate control of the land, often by temporarily leasing land and milk quota to their larger or younger, more ambitious neighbours, as well as securing other sources of income.

This is the most common feature of the European agrarian structure, 'whether we look upon worker-peasants as a successful adaptation to new conditions or an uneasy compromise forced on the farming population by declining living standards'.[28] Using their 'strategic reasoning',[29] they feel safer when they possess some financial assets to see them through hard times. Irish urbanites often find it difficult to understand that the very public practice of selling off a land parcel is only considered as an option by farmers in times of direst need, as a last resort. The preservation of the privacy of one's financial status is an under-researched, but nevertheless core constituent of farming culture in rural Ireland, and could hardly be underestimated as an explanation for some apparently illogical attitudes and actions.

Smaller farmers and/or those in peripheral regions are finding it more difficult to make ends meet on farming alone. Agriculture is declining in importance in Irish economic life and there is a general air of pessimism regarding its future. In the Irish state in 1973, 70 per cent of income was earned from farming in farm households. By 1987, this proportion had reduced to only 54 per cent, with 28 per cent coming from other direct income and 18 per cent from state transfers. This varied predictably by region, with 69 per cent being earned from farming in the south-east region and only 31 per cent in the north-west region at that time.[30] Despite the fact that this picture matches the EU ideal of rural people performing so-called 'multi-functional roles', what it means to most people in this position is financial struggle and long hours of hard work. Part-time farming is extremely common now. With regard to pluriactive farm households, Ireland is in fourth place in the EU.[31] In 2002, the farmer and/or spouse held an off-farm job on 48 per cent of all farms. This figure had increased by 3 per cent in just one year. On 35 per cent of all farms (up 2 per cent in one year), the job is held by the farmer, who is most likely to be in drystock farming, and demographically, more likely to be unmarried than full-time farmers. For a description of their incomes, see Table 3 below.

Table 3: Family Farm Income (FFI) for Full- and Part-Time Farmers

FFI (€)	% all Farms	% Farmers with Off-Farm Job	% Farmers no Off-Farm Job
< 6500	39	17	22
6500–13,000	22	10	12
13,000–25,000	19	6	13
>25,000	20	2	18
Total	100	35	65

Source: Teagasc National Farm Survey, 2002.

This table clearly shows that it is those in the low-income bracket who are most likely to take up an off-farm job. Of the 35 per cent with an off-farm job, 27 per cent have an FFI of under €13,000 p.a., so it is clear why they need more income. These farmers may do some contracting with machinery they possess, or they may work in construction, or farm relief services, or local industry. This is the option that makes most sense to small farmers, because it involves the least risk financially, and they can also maintain their links to the land and locality. It may also widen their social lives and hence improve their marriage prospects, giving them more confidence because of their increased income and new standing in the community. It therefore may mitigate against the worst symptoms of rural poverty and social isolation.

It remains an unanswered question as to whether this situation will continue for generations or whether it is a strategy adopted by those who are on their way out of farming anyway. There seems to be a general consensus at state level that part-time farming is a good thing for the Irish countryside, increasing incomes and preventing migration. The economist Jim Power disagrees with this position. He predicts that 'if we continue to encourage farmers down the road towards part-time farming, the Irish agriculture industry, as a competitive force, will be dead within twenty years'.[32] He believes that it is a means of hiding behind the fact that full-time farming is no longer viable, that the growth in part-time farmers will reduce the efficiency and competitiveness of Irish farming, and that it will weaken the political clout of the farming lobby. These statements led to his being reprimanded by then Minister for Agriculture, Joe Walsh. Notwithstanding this, his conclusion was that farmers 'need to wake up to reality that they may not be viable in the future. Holding on to a job and the farm will only lead to reducing quality of life'. These are harsh comments indeed, considering that combining off-farm work with farming is the path of least resistance for many. However, it is hard to imagine any other profession being advised that their best option for viability is to take another job altogether.

When they have to do a full week's work, farmers find it next to impossible to keep abreast of developments in farming and engage in the necessary investments for upgrading their farms. They may lose interest and let the farm work 'tick over', while the measurable week's work (and week's pay) takes precedence. From their employers' perspective, they serve as a valuable addition to the rural workforce. They are likely to work very hard for substantially less pay than their urban counterparts. Hence the prospects for exploitation may be quite high in an environment where union involvement would be completely unheard of. Working off-farm probably prevents them from considering innovations like organic farming, which could concurrently meet the needs of the changing marketplace, improve the long-term viability of farmers, and contribute positively to the maintenance of the rural environment. The countryside culture also suffers at a more general level, because these farmers retreat from defining themselves as farmers and hence, for example, are not as likely to participate in events like annual agricultural shows, which are vital for community bonding. This may lead to an identity crisis both at the level of the individual and of the group. These hybrid farmer-workers may welcome the economic capital that accrues from their off-farm jobs, and yet maintain a certain pride in their continued connection with the land. Of course, these sentiments may ebb and flow, depending on the ideological climate of the time and of the place.

Smaller farmers have been shown to be very slow to take risks and to become fully integrated in commercial markets unless they are forced to do so, because they understandably feel more confident in staying with the type of small farming methods that they know best. In order to retain their sense of independence, they are often willing to accept lower economic rewards than they could earn elsewhere.[33] The modernization perspective has long viewed this 'slowness' as a cultural condition. However, this reticence could be viewed simply as economically based logic where one simply weighs up the gains and losses that might result from an investment. It is not that they are averse to change and continue with familiar methods just for the sake of it, but rather that they may have very rationally opted for the certainty of the proven path over the risks associated with the unproven one. When scarce resources exist, as is the case for smaller farmers, this has to be a very finely tuned calculation because their very survival is at stake. Ortiz explains the nature of risk like this:

> When any individual, whether peasant, Western businessman or farmer, becomes aware that a particular action entails a risk, he evaluates it and decides how to act. The number of elements are numerous and relate to his income, cost of the enterprise, the payoffs, the type of assets he holds, whether his livelihood may be threatened, the alternative opportunities he foregoes, means of insuring the well-being of his family, his social status, etc. If on occasion a peasant hesitates to take a risk this does not imply that he always shies away from a gamble.[34]

Smaller farmers will invest in a new initiative, naturally enough, only when there is a high level of confidence that it will not threaten their subsistence. They do not have the resources necessary for the substantial investment required for commercial farming. Most often, there is a high level of self-exploitation, with people working several jobs and keeping all of their options open. Larger farmers, on the other hand, have a higher risk threshold, and are better endowed with the educational and financial resources needed to branch out into something new.

Cultural Capital

In the agricultural arena cultural capital, that form of social power that stems primarily from education, is almost as concentrated among a minority of farmers as is economic capital. Firstly, that group of large farmers who are responsible for the bulk of food production, who have absorbed the modernization message of the EU-influenced Irish state, and who have developed close economic and political alliances with the food industry, have become a farming élite. Intensive farming methods and high levels of biochemical inputs form the *modus operandi* of this group. They are termed progressive, viability or development farmers by the state. These large farmers possess cultural capital in abundance. They have accumulated it because their heirs attend agricultural college, learning the ethos and methods of productivist agricultural science. This is encouraged by the state because it was compulsory to have completed an agricultural training course to receive many of the EU grants that are available. This amounts to a closed circle, which ensures that one's chances of survival are tightly bound up with a particular agri-scientific educational ethos.

On the other hand, smaller farmers possess only small amounts of cultural capital. They are at the bottom of the social hierarchy in agriculture because they use extensive farming methods and have not adopted the productivism of the larger intensive farmer. Whether this is because they do not have enough resources to compete under capitalism, or whether they constitute a completely separate type of peasant producer is the crux of the difference between the Marxian and Chayanovian positions in rural sociology, as already mentioned. In his research in west Cork, Eipper (1986) adopts the latter position, arguing that these farmers are simply 'survival-oriented' rather than capitalist-oriented. They have received little formal agricultural training and they usually adopt the methods passed on as common-sense by their forebears and peers. The type of knowledge possessed by these farmers, for whom life and work on their farm is not a means to an end, but an end in itself, is viewed as backward and a hindrance by the dominant productivist ethos of the state.

At this point, let us consider one of the main coping strategies that the habitus of smaller farmers can accommodate – economic reliance upon the complete family unit. If structural factors are against them, then perhaps they can at least find strength in numbers. The family farm, of course, has had major ideological importance in Irish culture. Its magnitude is even inscribed in the 1937 Constitution in Article 45.2v, which states:

> The State shall, in particular, direct its policy towards securing that there may be established on the land in economic security as many families as in the circumstances shall be practicable.

This preference for small-scale production and the attendant apparent virtuousness of rural living formed part of the state's Catholic, ascetic, protectionist, anti-colonial, socio-economic vision after gaining independence. It was argued, for example, in the benchmark anthropological study of 1930s rural Ireland that sexuality only served the function of maintaining family status and that people's behaviour was very much circumscribed by traditional mores.[35] Since the modernization era began in the late fifties, and especially since EEC membership, this ideal has been largely consigned to the vaults of history. Productivist agriculture has been favoured, which necessitates larger and fewer production units, and, ironically, the 'family' has often been ripped apart by forced emigration.

The demographic profile of small farming households is an important consideration here. It is a crucial determinant of the chances of experiencing poverty, as it is related to the prospect of an off-farm income and its capacity to reproduce itself in the long term. It has been found in research on poverty in rural Ireland that those living alone, the retired and the unemployed, are those most at risk of becoming poor.[36] It is a fruitful approach to view poverty as a dynamic process rather than a state. From this perspective, poverty is seen as a condition that is actively produced, experienced, reproduced and perhaps resisted by actors in a social structural context.[37] However, it is important to remember that there is an intimate connection between the structural factors shaping the lives of people in rural areas and the choices made by the people themselves. For example, the amount of education received and the marriage strategies adopted cannot be separated from a structural context. These choices are inscribed in the habitus of social actors. Jackson and Haase assert:

> Poor demography is a consequence rather than a cause of changes brought about by shifts in the balance of the political economy of the country and countryside.[38]

One cannot underestimate the importance of the family network to farmers' survival strategies. It is now common practice to discuss the demographic as well as economic viability of a farm. This shows that the presence or absence of an heir who will commit to taking the farm into the next generation is vitally important. At a time when there is very little optimism regarding the future of small farming, most farming parents will encourage their children to take up another type of education so that they do not have to depend upon farming alone. It has been shown that young people from small farming backgrounds are much better educated than their urban working-class counterparts.[39] The sons, and especially daughters, of small farmers have fared remarkably well in all professional sectors, no doubt due in no small part to the encouragement they receive from their parents. One might suggest that these days, the offsprings' contributions of the *cultural* capital that is the reward for their educational efforts serves as a contrast to the emigrants' contribution of *economic* capital in the past. The former do not usually contribute to the survival of the family farm because they acquire status outside of agriculture, whereas the latter did in fact contribute to farm survival. Emigration and migration to Irish cities of small farmers' offspring remains as an important feature of their lives, providing as they do a willing and able labour force for the industrial and service sectors in the urban centres of the world economy.[40] This can have both a positive and negative impact upon the survival of smaller farmers. It can mean that it reduces competition between siblings for the inheritance of the family farm, or it can mean that none of the children stay, leaving the parents with no interested heir. The chances of either of these happening depends upon a number of social, economic and personal factors. Therefore the types of cultural capital accumulated by these offspring are effective in other areas besides agriculture, and they only indirectly relate back to the world of their parents.

Family strategies have been shown to be vital for farmers' chances of survival. Hannan challenged the common perception in previous sociological and anthropological studies that traditional peasant societies were dominated by the extended patriarchal family and 'that modernization would essentially lead to individualization, to the increasing isolation of the nuclear family unit'.[41] He argued that on the contrary, it was those farmers who most effectively utilized their kinship networks and family ties who were ultimately the most economically successful. Those who remained unmarried and heirless were effectively wiped out. He ended on a very pessimistic note, stating that if farming were not combined with some other occupation, small scale subsistence farming would end for good. This did not occur quite to the extent that Hannan predicted because EU remittances have provided a major financial bulwark for the smaller farmer. The poorest, however, have been pushed to the wall by market forces.

The economic input of different family members has been a contributor to the survival of smaller farmers, both past and present. Historically, remittances from the family members who emigrated were a vital source of income for the poor in both urban and rural Ireland, and functioned as a form of welfare when no other existed.[42] The Baseline Reports of the Congested Districts Board have shown historians that in the poorest parts of the country, like west Mayo and west Donegal, the money brought in by seasonal migrants like the 'tattie hokers' (potato-pickers) often accounted for up to one-quarter of the cash income of the poor up until World War I.[43] Emigration was a social safety valve that both reduced the numbers struggling for scarce resources at home and contributed much-needed income earned from their labours abroad.

The resilience of the family farm can be partly explained by the generation of off-farm income by farm women, as well as their substantial unpaid labour in the home.[44] Farm women's off-farm incomes have been found to constitute a vital part of the typically pluriactive modern farming household. This work is often in the professional and clerical sectors, but may also be in the service or domestic sectors. Because farmers' incomes are so often unreliable and intermittent, their spouses' more regular incomes can actually keep the family out of poverty. Women also contribute to the family's stocks of cultural and symbolic capital by the status derived from their own off-farm paid work. This is augmented by their participation in community-related activities, and also by encouraging the children to do well in school and/or take up extra-curricular activities. These combined efforts can often raise the whole family's social status in the community, which can pay off in other ways. In other words, cultural capital may sometimes be converted into economic capital.

Increased mechanization of farming has intensified the gender division of labour on farms, meaning that women on larger farms are doing less farm work than in the past, so their activities are now more likely to be in the house, or off the farm altogether. Women are often the ones to tackle the mounting piles of paperwork, which is now so vital to income maintenance. The farmer himself may not be confident enough to do this himself, and in the absence of a woman, may have to pay a professional to do it for him. These days, at a time when farmers are encouraged to diversify their economic activities, the woman in the family is often the one to take the initiative to set up a small business like a bed and breakfast or other agri-tourism venture. This can help the family to bootstrap themselves into a viable situation. While women rarely inherit land from their fathers, 11 per cent of Irish farms are owned by women. That this is now acknowledged is evidenced in the fact that the IFA appointed an Equality Officer, Mary Carroll, in

2002. In a press interview, she stated that she wanted farming women to be taken seriously, and not just seen as the ones to make the tea and sandwiches for the working men.

The Irish Countrywomen's Association (ICA) is a much underestimated organization that has been a very important social outlet for thousands of Irish rural women. It was formed in 1910 and its primary aim was to provide 'a vibrant social life in the countryside as a bulwark against emigration'.[45] This was not very successful but it nevertheless was hugely important. It served the other important function of bringing Catholic and Protestant women together socially, serving to counteract the bigotry that lack of interaction and isolation between groups fosters. The ICA in the twenty-first century, while its ideology is not quite parallel to the dominant liberal, secular trend within feminism in Ireland and elsewhere in the West, continues to raise issues – like the treatment of women in Irish prisons and provision of healthcare for women – that might otherwise go unnoticed. Perhaps it is an example of a type of 'indigenous feminism' that is attractive to ordinary women and does not necessarily follow the well-trodden path of Western liberal feminism. It is quite likely that its role has been and continues to be at least as important as that of female representatives in government in women's everyday lives.

Symbolic Capital

The realm of symbolic capital, or the prestige one accumulates by associating with society's 'movers and shakers', is also dominated by a minority within Irish agriculture. Larger farmers have managed to translate their economic wealth into political power by the use of their own particular kind of symbolic capital. They have been able to join the ranks of the capitalist sector, representing par excellence the capitalist ideal. In Co. Meath, for example, one in six farmers own over 80 ha, combining dairy, beef and cereal production.[46] This group is very adept at using the political system to their own advantage. Their power can largely be attributed to the political muscle they have exercised through the IFA. The social contacts they

made in the top tier of Irish agriculture consolidated their position in a circular fashion, thus reproducing the economic and
political status quo.

Larger farmers became relatively removed from local government, transferring their energies instead to the wider arena
of lobbying national and international politics by the IFA.
Founded in 1955, the IFA 'helped raise the status of farming
from being merely a way of life to that of a business operated
by persons with experience of policy, organisational and business affairs'.[47] The reason for this strategic change by these
larger farmers was that joining the Common Market augmented
their economic power, while simultaneously eroding their political power. The type of symbolic capital they possessed was
effective only at a local level. Examples were the control of local
co-ops and marts, strong representation in local government
circles and in farming organizations, easy access to credit, and
expanding their dynasties into other business sectors.

As decisions about their futures were now being made in
Brussels, they had to find new political strategies. For example,
they protested loudly when quotas were introduced in the mid-
eighties that curtailed their production limits. Many of the
large dairy farmers responded to this restriction by buying up
or leasing milk quotas from all over the country, before the rule
was introduced that the land where the quota was located had
to be bought or leased along with it. This is an illustration of
the interplay of the three types of capital. They had the economic capital to afford to buy up the quotas, the cultural capital provided the educational know-how to manage their
expanding farm systems, and their symbolic capital helped to
shape EU and national farming policy to allow them to do this,
at least for a while, until prevented from doing so by the state.
They are now protesting loudly again, about the implications
of the Nitrates Directive, which has actually been in place since
1991, and not acted upon until 2004. As mentioned earlier,
even this group's enormous social status is now being challenged by the forces of globalization.

Smaller farmers have been totally marginalized in this arena.

The main farming lobby, the IFA, is much more likely to address the needs of large commercial farmers.[48] The IFA's main concern over the years has been the maintenance of high prices for milk, beef and cereals at EU level. The benefits that resulted from this lobbying did not 'trickle down' but remained largely in the top tier through the EU pricing system. Some farmers interviewed for this research were scathing about the IFA's political agenda, asserting that they are not interested in the needs of the smaller farmer, and only serve the interests of the large, intensive dairy sector:

> To my mind, the IFA are the Irish Friesian Association, they just look after the milk man, they're not too worried about the beef man ... [milk men] are so big now that they have the power.

> Sometimes it's the big fellas are making plans for the small fellas. I think there should be more small farmers involved in decision-making.

In 2004 there was a proposal within the IFA to allow non-farmers to join, broadening its 80,000-strong membership to 'individuals or families who are living in the countryside or have an interest in agriculture and the countryside'. This was jokingly termed the 'Croke Park proposal', referring to the idea of soccer and rugby being played in the GAA stronghold of Croke Park (Rule 42). Even though the new members would not be able to vote, it was rejected outright at the January 2004 AGM, reflecting the IFA's exclusiveness as well as their defensiveness. Shortly afterwards, Joe Rea, ex-IFA President, asked in a letter to the *Farmer's Journal*:

> Where will this end? What about a badger welfare campaign; pressing for planning permission to remove ditches; requiring slurry spreading without smells; no more once-off rural houses?[49]

Another letter-writing farmer (in the same issue) had a more positive view of the idea:

> Countryside pursuits, be it horse-riding, shooting, fishing or hill-walking have an enormous potential to bring substantial amounts of money into rural Ireland and the real challenge is to get as much of that money as possible into the pockets of farm families.

This view is clearly a minority one. At a time when farmers feel aggrieved with their lot, the IFA are certainly not ready to open their doors to a plethora of people whose interests may directly contradict their own.

If the agenda of small farmers is excluded by the main farming organizations then the only avenue open to them is the local paternalistic basis of Irish political culture. There has been a long history of clientelist politics in Ireland, where the main role of politicians has been to get something for the local area from the centre of power. Chubb deemed the main role of politicians as 'going around persecuting civil servants'.[50] This is now being replicated at European level, and the function of any Minister for Agriculture is reduced more or less to that of a broker between his constituents and his European patrons. Farming Ministers have enjoyed relative political autonomy both at EU and national levels over the years.[51]

A recent example of the interface between localist and EU politics was the Kerry TD, Jackie Healy-Rae, lobbying at government level to have Kerry included in the Objective 1 region for the next round of structural funds. This type of political wrangling backfired in this case because the designation was ultimately perceived at EU level to be too ambiguous and the whole plan was rejected. In the local press, Healy-Rae attributed this to the following:

> I think that happened because of the unwarranted publicity that was created. It looked to the fellows in Brussels that they were being dictated to by people like me.

This is an example of a politician purporting to represent the relatively powerless, and attempting to get around the normal course of politics. It was an effort by the relatively weak local politician to make the system work to his own ends. Of course, it was a singularly unsuccessful attempt, but in clientelistic political culture, his constituents would greatly respect him for trying. The admiration that this TD has earned from the local electorate is now being transferred to the next generation of Healy-Raes, who are also throwing their hats (or caps!) into the political ring. This example is relevant because it shows that this type of clientelistic politics is still important in the social world in which small farmers live. The fact that this type of political strategy is so atomized and individualistic leads us to the question of class identity and what potential exists for collective action on that basis.

It is argued here that farmers cannot be said to possess a unitary class identity, because there are as many factors to divide them as there are to unite them. Their class identity can often be seen as contradictory because they may occupy more than one class position at once, that of bourgeois landowner, urban proletarian, whose wife and/or children may work in the service sector or the professions. Class, when measured in this rather minimalist fashion, is not very useful for examining the social dynamics of the countryside. However, the form of theorizing that emphasizes lived experience over material interests brings class to life. E.P. Thompson, for example, viewed class as a dynamic *happening* rather than a static *thing*. He asserts:

> Class happens when some men, as a result of common experiences (inherited or shared), feel and articulate the identity of their interests as between themselves, and as against other men whose interests are different from (and usually opposed to) theirs.[52]

Building upon Thompson's contribution, we can alternatively begin with farmers' actions as class collectivities. They are often divided because different groups of farmers might strive towards divergent aims, e.g. EU support for various sectors within farming, like dairy and arable production. They can also be divided according to size of farm, or region of the country, or of which farming organization one is a member. Small farmers may be lobbying for advantages that accrue mostly to the top tier of farmers. The likelihood of demonstrating a united class-based front is probably quite limited, especially when they couch their protests in the rhetorical language of rights. The state, and indeed the public, probably no longer feel that farmers have an automatic right to financial support. This approach is not very meaningful in Irish public culture, because farming is fast losing its cultural significance in Ireland as a symbol of authenticity and tradition.

Protesting on the streets in tractor cavalcades antagonizes the public rather than attracts their sympathy. The only way for smaller farmers to gain more popularity at this point is to take environmental issues seriously and adapt their practices to consumer demands. The IFA cannot be relied upon to represent their interests, and the EU and the state will only subsidize them for eco-friendly farming practices. Many of them may lack the know-how to deal with this situation, but this needs to be rectified if they want to get off the treadmill of simply biding time until their eventual disappearance from rural life.

Conclusion

The process of globalization, with its rationalizing impact on agriculture, is often blamed for the demise of small farmers. However, farmers differ in their abilities to cope with these pressures. While such global power structures do have a major impact on people's lives, this is not the complete story. Long states:

> It is theoretically unsatisfactory to base one's analysis on the concept of external determination. All forms of external intervention necessarily enter the existing life-worlds of the individuals and social groups affected, and in this way are mediated and transformed by these same actors and structures.[53]

This chapter has sought to put a human face on the statistical decline of farming in Ireland. This decline is far from uniform, and those who manage to make a decent living from the land do so because they possess certain sets of socio-cultural resources, or capitals. There is evidence that something like an élite exists, who have done extraordinarily well from EU accession. The mechanisms of the CAP have disproportionately enriched the coffers of larger farmers in Ireland and contributed to their upward social mobility. The majority just muddle through, however, making the most of their circumstances and using what assets they can to ensure their survival. This is not a random process, however, as there are discernable social trends. For example, those that cannot survive on their farm income usually hold on to the land if they can. It is argued that people's everyday actions have an intimate relationship with the underlying social structures that determine their power positions.

The picture of Irish agriculture portrayed here is one of gross inequality. The modernization project pursued by the Irish state has relegated all but a minority of farmers to the social fringes. The intensification and commercialization of Irish farming continues apace, and we have seen how Bourdieu's theory of practice, and his concepts of habitus and capitals, can help us to develop an alternative mode of analysing farmers' coping strategies in the face of this increasing commercialization. The next chapter deals with the reform of the CAP, which paved the way for the introduction of environmental concerns into the agricultural field.

chapter 3

Changing the Goalposts:
Why the CAP Needed Reform

It has been shown that the CAP ultimately widened the gap between richer and poorer farmers. It awarded those who produced the most goods and pushed smaller farmers to the social margins. It fostered an atmosphere of greed and encouraged practices that were both socially and environmentally devastating. The following chapter will look at the political context of the reform of the CAP from the mid-eighties to the present. Farmers have had to adjust to a whole new set of requirements and make the most of whatever the EU had to offer. No other sector has to cope with the amount of change that farmers do. It is argued that the ultimate effect of CAP reform so far has been the zoning of the Irish countryside into productive and environmental zones, divided predictably along east/west lines. The social effects of the new developments in EU policy are more difficult to predict, however, as farm payments are decoupled completely from production.

The First Round of CAP Reform: Budgetary Concerns
To this day, productivist agriculture continues more or less unabated in the EU. It has nevertheless been questioned since the mid-eighties because of the perceived economic inefficiency,

environmental destructiveness and social injustice that it has caused in rural Europe. However, there is evidence that this realization dawned a long time ago. A 1980 European Commission report recognized the growing tendency both toward surplus production and social inequality,[1] and proposed making individual countries more accountable for their actions in the agricultural arena. However, it was several years before action was taken on these concerns. When asked why he thought CAP reform took so long, a DAF civil servant responded:

> Political change is slow. In order to convince people of the need for change, you have to create the right atmosphere in order to get people on side. To get a political decision passed in the Commission, you can't go in like a bulldozer or you'll encounter opposition. You have to do your homework first. If a suggestion is defeated, it's a lot harder to get it off the back-burner in future.

In the mid-eighties, it was recognized that there were more farmers than were needed to produce enough food for the EC, so the definitions of the uses of the countryside had to be altered. The land was not supporting as many people as it had in the past and was also being used in many different ways. It could also be argued that farming as a profession had been losing, and continues to lose, respect in public discourse. Many people began to think that they had been usurping too much of Europe's resources and that the urban European taxpayer had spent long enough subsidizing them. Agriculture no longer enjoyed the monetary and ideological dominance that it had in the past, and pleas for solidarity with the 'country cousins' often fell on deaf ears.

In some countries, notably Ireland and France, peasants or small farmers had previously been constructed by the state as 'the cement of the nation'.[2] However, this ideology of what is termed 'rural fundamentalism' was broken down by the growth of a wealthy farming class who were, as we have seen, virtually created by the EU pricing mechanisms. The sympathy that farmers formerly attracted wore very thin because the more capitalist they became, the further they retreated from the idealized peasant of nationalist ideology. When farmers began to drive more Pajero Jeeps than horses and carts, the love affair between the nation and the farmer grew cold. The picture-post-card images of rural Ireland became more and more divorced from reality. New and competing claims were made for the best type of future for the European countryside. This coincided with a significant trend of counter-urbanization in rural Europe, including Ireland. This refers to people moving out of cities and rediscovering the joys of country living. Some pockets of the Irish countryside, such as west Cork, are attracting former urban-dwellers and buyers of second homes in the regions. Many other rural communities are becoming increasingly isolated, however, as services such as schools, post offices, garda stations, and shops are being closed down.[3]

This change of emphasis produced the impetus for the reform of the CAP. Other factors leading to the CAP reform were:

- the burden of the high cost of storing food surpluses
- political dissatisfaction with the high consumer costs for food
- the international political tensions that resulted from the EC's financial support of its exports[4]

The first of these factors, the expense of storing and disposing of the food surpluses generated by EC protectionism and productivism prior to the mid-eighties, was probably the key factor in this seachange.[5] This is verified by an EC report from 1987, which states:

> the Community bodies have thrown themselves into the task of restructuring the CAP, *which has been made urgent by the accumulation of surpluses and the critical budget situation*[6] [my emphasis]

The presence of new social groups also created new sets of demands on an increasingly diverse countryside. This era has been termed 'post-productionist' or 'post-productivist' by some authors.[7] However, this may not be a precise account of this phase. When interviewed, a DAF official asserted:

> The first round was more market-driven than anything else, as a way of coping with over-production. Concern for the environment didn't come into it at that stage ... The economy came first, before people or the environment.

This suggests that budgetary issues took precedence over a more fundamental questioning of the EC's productivist mission.

By the mid-eighties, the CAP was absorbing 70 per cent of the EC's budget, which was completely disproportionate to this sector's relative economic importance to the EC.[8] High food prices paid by the consumer and remittances earned from import levies paid for the protection of EC farmers from international price fluctuations and the storage and periodic export of surpluses.[9] This also resulted in enormous injustices committed against the developing world, as their products were discriminated against upon import into the EC. Productivist EC policy had encouraged unprecedented levels of investment by farmers and hugely increased farm productivity, which led to the overproduction of many commodities such as cereals, beef, milk and wine by the late seventies, and the transformation of the EC from a net importer to a net exporter of food. In 1982/3, the self-sufficiency rates in sugar, wine, cereals, milk products and meats were 147 per cent, 125 per cent, 117 per cent, 118 per cent and 100 per cent respectively.[10] In the mid-eighties, there was a surplus of 16.8 million tonnes of grain, 1.5 million tonnes of butter, 1.1 million tonnes of milk powder, 0.6 million tonnes of beef and 5 million hectolitres of wine.[11] By 1987, the EC had an estimated ECU12.3 billion worth of food in storage but deterioration meant that its real value was only one-third of that amount. It is hard to forget Bob Geldof's meeting with an unsympathetic Margaret Thatcher in 1985, where he made a memorable plea for these surpluses to be sent to those who were starving in Ethiopia at the time.

Taking all of this into account, an EC report from 1985 understandably said that:

> The Community can no longer afford to allow EAGGF expenditure to rise unchecked as it has done in the past, by 25–30 per cent between 1975 and 1980.[12]

The increase of these surpluses led to a drop in the price paid to farmers for goods produced. For example, the world market price for cereals and oilseeds exported by the EC fell by 40 per cent between 1980 and 1987.[13] This increased pressure on farmers to produce more in order to stay in business. The EC had to reduce the production of the goods that were already in surplus, in order to render farming more efficient. This created a very uncertain economic climate for European farmers, who suspected that the good times were coming to an end.

The amount of EC money going to farming had to be reduced, despite lobbying from the farming organizations and individual member states' Agriculture Ministers, some of whom were more effective than others.[14] There was frustration among the environmental groups and the public at large about the power exerted by élite groups who sought to obstruct meaningful CAP reform.[15] The main function of CAP reform was therefore to cut the cost of maintaining farming, thus reducing the burden on the European taxpayer, who 'was no longer prepared to underwrite capitalist accumulation in the agro-food system'.[16] It had become clear to EC policy-makers that the legitimacy crisis from which agriculture was now beginning to suffer had to be addressed. Various conflicting factions surfaced around the issue of the EC's EAGGF budget. Three such factions were:

- larger farmers in advantaged regions who sought to prevent major change occurring in the price policy
- smaller farmers in more disadvantaged areas who sought to improve their own position with aid from the compensatory policy
- those who sought the redirection of funding from agriculture to other sectors[17]

A DAF civil servant told me that the farming organizations, for example, 'looked at how CAP reform would impact on themselves rather than the bigger picture'. This 'bigger picture' was the international political stage, which cannot be ignored. The CAP had become a major problem in the EC's input to the GATT[18] negotiations, and the US began to demand its reform. The protectionist bias of the CAP was 'sharply at odds with the US goal of complete liberalisation by the year 2000'.[19] The result of all of these negotiations was a 'conservative change'[20] because the Guarantee Section (or the pricing mechanism) of the EAGGF remained the primary means of supporting agriculture, but yet it did place some limits on agricultural production for the first time since the formation of the CAP.

Preservation of the environment was also becoming a concern at this time. It began to be recognized by policy-makers that many of the practices involved in the industrial model of rural resource use had been seriously harming the environment. This awareness of the environmental ill-effects of capitalist accumulation and exploitative land use constituted 'a major cultural shift'.[21] A DAF official commented:

> I suppose you could say that a lot of the new changes in policy were an attempt to redress the imbalances caused by the Commission in the past. But it's easy to have 20/20 vision in retrospect.

For example, the intensification of farming had led to increased dependence upon chemicals and heavy farm machinery. The use of technological and biotechnological inputs increased by 60 per cent in the twenty years between 1966 and 1986, and the consumption of nitrogenous fertilizers almost doubled in the same period.[22] This growing use of environmentally aggressive farming methods led to the undermining of farm animals' welfare, destruction of bird and animal habitats, water pollution, pesticide residues in food and soil erosion. The environmental costs of productivist agriculture, such as the outright extinction of many plant and animal species, had been externalized, or in other words, borne by the broader society.[23]

While a 1985 European Commission Green Paper (commonly called the Green Book) proposed a decoupling (or separation) of commodity prices from farm income and rural development objectives, powerful lobbies representing agribusiness interests vetoed such radical reform for several years, until the early nineties.[24] As a result of these protests by powerful agro-food lobbies, who were the main beneficiaries of the CAP, these concerns were set aside in favour of a strategy based on controlling rates of production, rather than seriously addressing environmental concerns. This first round of CAP reform was a benchmark policy initiative with the following aims:

- quotas being placed on many major agricultural products: mainly, of course, milk

- levies being placed on all surplus production
- more funds going to disadvantaged areas in the form of headage payments
- new measures being introduced to encourage farmers to protect the environment (expanded in the second round of CAP reform in 1992)
- introduction of an enhanced farm retirement package
- introduction of incentives to encourage diversification e.g. into forestry[25]

However, as this round was primarily concerned with cutting down on the production of surplus goods, the environmental content remained rather underdeveloped here, but became much more evident in the second round of CAP reform,[26] as we will see in the following section.

The Second Round of CAP Reform: Environmentalization

Following these changes, which were brought about by Ray MacSharry, the then EU Agricultural Commissioner, it became evident that even more regulation was required and a second round of reform was initiated. Those designing the 1992 CAP reform were faced with handling and ameliorating the social injustice and environmental destruction that had become the norm both in Irish and broader European agriculture. The Maastricht Treaty also extended the EU's range of concerns to incorporate industry, science, technology, the environment, security and defence, social security and education. The farming lobby was therefore formally overtaken by the environmentalist one, among others. Overall, it constituted an attempt to delink the types of support that went to farmers from the market mechanism.

EU agri-environmental policy resulted from this second round of CAP reform in the early nineties, which in turn resulted from wider international developments in agricultural restructuring. A key feature of this era was the environmentalization of agricultural policy. One DAF official, interviewed in the course of this research, attributes the fact that environmental concerns began to penetrate all policy spheres at this time to three factors:

- the public had become more sophisticated, more environmentally aware, and more aware of their rights in terms of access to information
- more scientific research had been conducted, producing concrete data that could help identify areas in which reforms were required
- fiscal policy dictated that more accountability and efficiency was needed within the Commission, and it was no longer acceptable that one policy would cancel out the impact of another

He went on to say that politicians and policy-makers 'could no longer hide behind authority and had to act more responsibly than heretofore'. This restructuring of agricultural policy has been explained as being motivated by

> the need for the state to strike a new balance between the continuing requirements of capital accumulation in agriculture and the necessity of maintaining the social conditions – minimum levels of political consensus and quiescence – under which accumulation in general can occur.[27]

The implications of international processes of economic and social restructuring for individual countries has been highlighted as a key area for analysis.[28] It is important to determine the real effects of global trends on local actors, based as they are in particular localities. Funding had to be cut for production, and simultaneously an alternative source of income had to be provided for those who could not compete in an increasingly deregulated market, in order to preserve the fabric of rural society. Environmentalism was used as a means of framing these goals in a politically acceptable manner. It is argued later that REPS can be viewed as one key policy initiative that sought to address these problems in Irish agriculture. One agricultural economist provides us with a determinedly unromantic statement of the underlying economic thinking behind agricultural restructuring when he states:

> A sentimental attachment to the idea of the 'family farm' (whatever that might be) and a reluctance to abandon the land to the wild are flimsy justifications for the enormous expenditure involved [in supporting agriculture]. *The redirection of significant areas of land to forestry and manmade wilderness might prove more appropriate for the needs of the late twentieth century.*[29] [my emphasis]

Agricultural policy-makers from this point onwards were less concerned with the needs of small farmers trying to survive in the occupation of their choice in marginal or peripheral areas, and more concerned with the recreational needs of the urban population. The EU, from this cold economistic perspective, needs a much smaller number of farmers than it does at present. New functions had to be created for the land, and for those working it, in light of the over-production of food. As early as 1990, a pattern of land use was being established throughout the EU, including Ireland, 'whereby a category of productive farms co-exists with a growing proportion of holdings that must be "allocated other roles" as "resource managers" in the rural economy'.[30] This was a prophetic statement indeed, in light of subsequent developments.

There is now a stated policy concern about the transition from the productivist and protectionist model to open competition on the world market, and a new concern for food security, environmental conservation and social welfare in countryside.[31] The most potent image in EU rural development policy is that of the 'multi-functional countryside', which should be dedicated to the 'production of renewable raw material for non-food purposes or the energy sector, rural tourism, marketing of high-quality produce or the preservation of our cultural heritage'.[32] This sanitized and idealistic terminology belies the fact that farming is not quite like any other industry, where workers can simply be entreated to seek alternative work outside of an unprofitable sector. Firstly, there is usually a lot at stake emotionally and also in terms of maintaining social status in the community. Secondly, the land they own is to an arguable extent a public resource, or 'common good', and it affects everybody if it is not maintained, and left to return to its natural state. Therefore, an alternative is for the majority of farmers to use their land for the production of 'environmental goods', and let the most intensive and efficient large farmers produce the 'food goods' at lower cost. In this way, vast portions of the countryside can become the green lungs for urban society. Arguing from within a British context, Ward suggests that the designation of some areas as environmentally sensitive, and others as centres of efficient agricultural production, could produce a 'two-track countryside'.[33] In the Irish context rural space, and the people who occupy it, is also being zoned into agricultural and environmental uses.

How then was it possible to cut agricultural funding in a protectionist system like the EU? To rely on the market pricing mechanism to exert control over goods produced may not be enough. Many agricultural commodities are insensitive to price because there is a long production cycle for such goods. Also, farmers do not always behave like a profit-maximizing *homo economicus*, and may not wish to change their farming practices. Results from this are therefore probably more visible in the long term than the short term. In terms of a managerial strategy, limits may be placed either on agricultural outputs or inputs.[34]

(A) Limits are placed on the volume of outputs through such policies as set-aside. In Ireland since the 1992 reforms, the farmer cannot use 10–15 per cent of his land from 15 January to 1 September. Outside of this time, it can be used for grass for the grazing of the farmer's own stock only. However, it has been the experience elsewhere in Europe that farmers usually set aside their least productive land, therefore production is not in fact cut back by very much. It appears that set-aside has had limited success in satisfying the minimalist EC objective of controlling the supply of agricultural goods to the market. It is thought that farmers compensate for setting aside this sizeable proportion of their arable land by applying increased levels of fertilizers and pesticides on their remaining productive land. It therefore may have had the effect of increasing rather than decreasing the productivity of farming land in the EU. The effects of this resulting intensification on wildlife and plant biodiversity are not actually proven, but one can only surmise that they will probably have been negative. It has also been pointed out that farmers whose production is already limited by set-aside will be extremely reluctant to allocate even more land to agri-environmental programmes, or if they do so, they would need substantially higher payments than currently exist.[35] The designation of certain areas as environmentally sensitive, and thus 'untouchable' by farmers, is also part of this strategy and will be discussed in depth in chapter five.

(B) Agricultural funding can also be cut by limits being placed on amounts of agricultural inputs. This approach involves cutting back on the consumption of biochemical inputs such as fertilizer, which are used to enhance production levels and, at a broader level, encouraging farmers to become less intensive. This policy endorses the uptake of ecologically friendly farming methods among farmers. This is also compatible with the aspirations of those seeking to further develop rural tourism. Agri-environmental policy in Ireland, therefore, has been the product of intensive negotiations between the various bodies that have an interest in farming and the environment. These include representatives of EU states, the EU Commission and the environmentalist lobby.

While the foundation was laid earlier, environmentalists have had quite an impact upon European agriculture since the early nineties. One Irish Green Party member attributes the rise of environmentalist concerns to the fact that after the fall of the Berlin Wall, it was clear that the environment was damaged by both capitalism and socialism in both western and eastern Europe:

> they both ended up on the same road of worsening air quality, water quality, soil erosion and that type of thing and that there had to be a more community-centred, more planet-centred approach to a better society.

The environmental lobby in Europe professed the same message as those who were concerned about the waste of the EU budget. Environmentalists advocated cutting back production, which helped with the over-spending problem. This cost-cutting produced a steady decrease in the amount of money spent on supporting farming. The proportion of overall EU expenditure on the guarantee section of EAGGF dropped from 64.2 per cent in 1988 to 50.5 per cent in 1996.[36] Lowe argues that this change came about 'not necessarily through any deep convictions', but rather as a cost-cutting exercise for the CAP.[37]

This resulted in an unusual (temporary and opportunist)

alliance between proponents of ecology and advocates of productivist agriculture. Because the latter were so powerful politically and could not realistically be confronted by the Green
lobbyists, a political compromise was reached, whereby agri-
environmental policy was imposed upon those farmers who
were relatively powerless, namely those at the smaller end of
the scale and those living in peripheral areas like the west of
Ireland. So while environmentalists and agronomists would
usually have contradictory aims, in fact they both share two
major characteristics: both rely on the cultural capital of high
levels of scientific education and expertise, and neither group
trusts small farmers to manage their own land. Small farmers
are excluded from both the conservationism of the former and
the utilitarianism of the latter. The strategy, therefore, of
compartmentalizing the land into zones designated for environmental protection and commercial farming suits both
groups equally well. Each would place higher value on their
own preferred type of land use, but each also tolerates the existence of the other. This gave rise to the new policy arena of
agri-environmental policy.

Agri-Environmental Policy

The most important regulation that legislates for agri-environmental policy is EU Reg. 2078/92. The political compromise
that this regulation entails became possible following the ele-
vation of environmental protection on the EC agenda by the
passing of the Single European Act in 1987.[38] It formed part of
Ray MacSharry's approach to CAP reform in the early nineties.
It recognized that it was neither economically or politically fea-
sible to continue to support the unprecedented growth in agricultural production. Aid schemes that derive from this regulation are financed by the guarantee section rather than guidance
sections (or price rather than structural) of EAGGF funding,
thus bringing environmental concerns into mainstream policy
and allocating much more funding to alleviating them than previously.[39] This proceeded faster in Europe than in the US because

the European Parliament has revealed itself to be more sensitive to the desires of consumers and taxpayers than has the US government.[40] Consumer concerns over food quality were also an important motivating factor for agricultural policy formulation at that time.[41]

EU Reg. 2078/92 was an adjunct to the previous regulations on set-aside, extensification and protection of environmental zones that had been enacted throughout the eighties, but which had not been taken seriously by member states. For example, the precursor to 2078/92 was Article 19 of EU Reg. 797/85, which was a voluntary scheme that established the designation of Environmentally Sensitive Areas. This attempt had limited impact because it was only taken up in the northern member states, being ignored by the economically poorer southern states and Ireland until c.1990.[42] The difference with this measure was that it was compulsory and could not be ignored.

Ecology is central to the *raison d'être* of EU Reg. 2078/92. This emerged in tandem with two other important voluntary regulations, EU Regs. 2079/92 and 2080/92. The former legislated for early retirement, while the latter legislated for a forestry aid scheme. EU Reg. 2078/92 is an aid scheme that aims 'to encourage agricultural production methods compatible with the requirements of the protection of the environment and the maintenance of the countryside'. This managerial approach aims to provide aid for farmers who undertake:

- to reduce substantially their use of fertilizer and/or plant protection products, or to keep to the reductions already made, or to introduce or continue with organic farming methods
- to change to more extensive forms of crop, including forage, production, or to maintain extensive production methods introduced in the past, or to convert arable land into extensive grassland
- to reduce the proportion of sheep and cattle per forage area
- to use other farming practices compatible with the requirements of protection of the environment and natural resources, as well as maintenance of the countryside and the landscape, or to rear animals of local breeds in danger of extinction
- to ensure the upkeep of abandoned farmland or woodlands
- to set aside farmland for at least twenty years with a view to its use for purposes connected with the environment, in particular for

the establishment of biotope reserves or natural parks or for the protection of hydrological systems
• to manage land for public access and leisure activities[43]

EU Reg. 2078/92 sets out the structure of the aid scheme, the amounts of money involved, and the conditions that had to be implemented. It was to be monitored regularly to ensure compliance. Member states themselves were to determine the terms and conditions of payment, and the amount of money paid. Such new regulations clearly indicate that some attempt was being made in the second round of CAP reform to integrate environmental concerns into European farming policy. This has led to funding being provided for agri-environmental schemes in all EU member states, of which REPS is the Irish version. These agri-environmental schemes have been viewed both positively and negatively since their inception in the early nineties. On the one hand, the Green Group of the European Parliament have been very critical of such schemes, saying that they 'will work only as a meagre fig-leaf to cover the consequences of an environmentally and socially destructive agricultural policy'.[44] On the other hand, a farmer from Cavan stated:

> REPS is a very good scheme. It benefits the farming family, the farming environment and the rural community.[45]

The remainder of this chapter provides an analysis of the scheme that goes beyond these sharply opposing views. It is argued that farmers have been allocated the roles either of commercial food producers or environmental managers. This casts an environmental role for uncompetitive farmers, which is legitimate in the public eye and is politically popular at a time when farming has been losing public credibility. This ultimately has the effect of leaving the existing productivist regime intact.

REPS: The Details
REPS is the main agri-environmental scheme operating in Ireland, agreed in 1992 in Luxembourg, and initiated two years later. REPS I ran from 1994 to 1999, REPS II ran from 2000 to 2004, and REPS III began in the summer of 2004. In REPS,

farmers are encouraged to adopt new farming methods, to replace their previous apparent exploitation of the environment. In order for this system to work, stringent rules have to be followed. Farms are inspected on a regular basis and penalties are implemented in the event of any breaches in the rules. The following measures are explained in the very detailed Book of Specifications issued by the DAF. This was altered slightly with each new round of REPS, being first updated in 1999 and again in 2004. In REPS I, farmers were paid a basic rate of £50.67/acre, or €158.98/ha for keeping to these measures, with a maximum of 99 acres (40 ha). When interviewed, one civil servant in the REPS office stressed:

> In REPS, the average is about 37 ha, it is geared toward the smaller farmer because its harder for them to survive in farming ... we try to compensate farmers to keep people on the land and to keep them more viable.

This was therefore clearly designed as an income support for the smaller farmer. The maximum any farmer could earn in REPS I (outside of the supplementary measures) was £5000 (€6400) per year. The average REPS I participant had 92 acres (37 ha) and an average annual payment of £4600 (€5900). This by no means represents the smallest farmers, but rather medium-sized farmers. This group has the most to gain from REPS and the highest capacity to take risks with new ventures. This has also been found elsewhere:

> even within the classification of 'small cultivators', one finds peasants with somewhat more land and other means of economic survival, who, given their somewhat greater access to resources, are more likely to accept new innovations, because failure is less likely to result in destitution.[46]

REPS I did not have any impact upon the incomes or farming practices of the smallest farmers, who own under 50 acres (20 ha). Many of these did not join, and if they did, the income they received was fairly negligible against costs

The eleven basic REPS measures that have to be followed by farmers are as follows. These have not changed since 1994, so they have been the basis of REPS I, II and III.

1. Nutrient Management Plan: The aim of this measure is to minimize nutrient seepage from agriculture into water resources. Pollution in the form of nitrogen, phosphorous and lime must be avoided, with the adoption of a Nutrient Management Plan. This is clearly the most important measure, with the longest list of detailed instructions for farmers.

2. Grassland Management Plan: This is designed to produce sustainable grass cultivation that minimizes poaching, overgrazing and soil erosion, and to protect habitats. This plan must be compatible with Measure One. The stocking level on the farm, the extent of already existing degradation, and housing arrangements for cattle are therefore key elements.

3. Protection and Maintenance of Watercourses and Wells: This is to improve water quality for fish and wildlife, and to improve the vegetation alongside rivers and streams. Under this measure, access to rivers by cattle and the application of herbicides, pesticides and fertilizers within 1.5 m of rivers or streams are banned.

4. Retaining Wildlife Habitats: This aims to protect natural features of the landscape such as woods, marshes, peatlands, eskers, hedges and old buildings. Commercial farming is to be curtailed in certain areas. Practices such as drainage, afforestation, turf-cutting and burning are all prohibited.

5. Maintenance of Farm and Field Boundaries: Boundary and roadside fences, stone walls and hedges are to be maintained in the interests of wildlife, stock control and scenic appearance. All outer boundaries of the farm must be stock-proofed and there must be hung gates in all roadside gaps.[47]

6. Use of Herbicides, Pesticides and Fertilizers Near Hedgerows, Ponds, Streams and Wells: The aim of this measure is to protect the flora and fauna that live in or near hedgerows, ponds and streams.

7. Preservation of Features of Historical and Archaeological Interest: This is to protect the historical monuments that are dotted all over the Irish countryside. The Office of Public Works' (OPW) maps of archaeological remains should first be consulted. These features should not be tampered with and animals should not be allowed access to them.

8. Maintenance and Improvement of the Visual Appearance of the Farm and Farmyard: This is to promote improved awareness of integrating the visual features of the farm and farmyard into the surrounding countryside, which should be kept clean, tidy and visually appealing.

9. Production of Tillage Crops Without Growth Regulators, Without Burning Straw and Stubble and leaving Field Margins Uncultivated: No plant growth regulators should be used, straw and stubble must not be burned and field margins should be left uncultivated for at least 1.5 m.

10.Training in Environmentally Friendly Farming Practices: Farmers are to be encouraged to attend the REPS training days and farm walks, for which they are paid £100 (€127).

11. Keeping Such Farm and Environmental Records as May Be Prescribed by the Minister: Farmers are to keep monthly records of everything they do on the farm, from buying and selling livestock to slurry application, to purchase of chemical fertilizers and anything else deemed relevant. REPS payments are dependent on the keeping of these records. This is viewed as quite a complex task by many farmers and many feel that they require expert help from farm advisors to keep up-to-date with these requirements.

11a. Supplementary Measures: There is extra pay for the farmers who want to go further than most in terms of environmental protection. Along with the eleven compulsory measures listed above, farmers can elect to sign up to supplementary measures.

These are as follows:

a) Natural Heritage Areas (NHAs) and b) Degraded Areas (commonage) – now called Supplementary Measure A: Having been separate in REPS I, these two measures were combined in REPS II and made compulsory. The SMA (Supplementary Measure A) covers NHAs, Special Areas of Conservation (SACs) and Special Protection Areas (SPAs) on farmland and all commonage. This would appear to be in line with the combination of all of these under NATURA 2000. The only source of income available to farmers who own designated land was now tied to REPS participation. This new REPS measure is to allay the understandable widespread confusion and fear over rules that apply in each of these types of protected areas and to provide an income for farmers living in those areas. The first 40 ha (100 acres) qualifies for a payment of €242/ha or £77.16/ac, the next 40 ha, €24/ha or £7.60/ac, the next 40 ha, up to the limit of 120 ha, €18/ha or £5.70/ac. This payment was in return for numerous controls and restrictions on farming practices. It was a compromise solution that followed protracted year-long negotiations between interested parties. It was only in August 1998 that it was agreed that farmers whose land is held entirely in commonage in Environmentally Sensitive Areas (ESAs) of the west are eligible for REPS. Previously they had to have over ten acres of privately held land. Also, REPS payments were restricted on commonage, adversely affecting the income of 8000 of the 10,000 hill sheep farmers in the country.

The requirements to qualify for these payments are to clear all sheep off large areas of overgrazed commonages along the Western seaboard and to join REPS. At a course given by the DAF and the Department of Arts, Heritage, Gaeltacht and the Islands (DAHGI) to potential REPS planners, the guideline was that where 5 per cent of commonage was overgrazed the recommended remedy was 80–100 per cent destocking. The amount of land in question is quite sizeable, the area under commonage in Galway being 117,000 ha, Mayo 100,000 ha, Donegal 99,000 ha, Kerry 83,000 ha and Cork 31,000 ha. It is now compulsory under cross-compliance rules for farmers in these areas to meet REPS specifications. Ewe premia and sheep headage payments are now tied to REPS, so they are effectively forced into this system. The EU budget provision for headage was cut from £119m in 1997 to £88m for 1998, and there was a corresponding 43 per cent increase in funding for REPS in the same year.

Other rules also applied, like a ban on turf-cutting in SACs. Such new rulings met with some opposition when first introduced. A meeting took place in Maam Cross, Co. Mayo in December 1997, where farmers complained bitterly that these were incompatible with Connemara farming. Paul Mooney reported that the farmers said that some of the SAC controls were 'nonsense' or 'impractical', like the rule that dosed cattle were to be kept off the land in SACs for seven 7 days. They framed their complaints in terms of losing ownership of their land and livelihood. Their right to do as they like on their own land is now circumscribed by EU rules and this is a bitter pill to swallow. Traditional methods of farming and the knowledge systems that underlie them are undermined by this. These farmers claim that soil erosion is caused as much by the weather as by sheep on the hillsides and that it is happening in areas of Donegal where there are no sheep.[48] It was even admitted at a national conference on REPS in January 1999 that REPS did not counteract overgrazing of commonages in Connemara at all since it was introduced in 1994. An expedition of EU inspectors recently examined the degraded commonages of the West and 'they're not at all happy', according to a senior DAF official. Whatever resistance they attempt, it makes no difference to the fact that commonage farmers' 1999 ewe quotas were cut by 30 per cent in counties Donegal, Leitrim, Sligo, Mayo, Galway and Kerry, as the fast-track to entitlement for REPS II, which was now their only source of income.

c) Local Breeds – Cattle, Horses, Sheep: This is to encourage farmers to breed, for example, Connemara ponies, Galway sheep, Irish Draught horses, and Kerry, Dexter and Irish Maol cattle. The farmer gets paid for each qualifying animal. Teagasc warn that the farmer needs a genuine interest in rare animal breeding, and there is a 100 per cent penalty for any breaches. Since the combination of NHAs and commonages under SMA, farmers in those areas are now also allowed to claim under this measure. This is the only exception to the rule that only one supplementary measure can be claimed. This would appear logical since the majority of these rare breeds originated in these areas of the country.

d) Long-Term Set Aside: An additional payment is made for land in riparian zones along rivers and lakes that is set aside for at least twenty years. This payment is subject to a maximum of 2.5 ha per farmer. Uptake of this measure has been minimal.

e) Public Access and Leisure Facilities: An additional payment is available where public access is allowed for environmentally friendly leisure activities. No more than 200–300 have signed up to this measure, and indeed this issue has caused much acrimony in some areas.

f) Organic Farming: For the first two years of conversion to organic production, the annual REPS payments are about €349/ha or £111/ac, the payments from REPS and IOFGA/Organic Trust. Once they are fully organic, the annual payments are €254/ha or £81/ac.

In REPS II, beginning in October 2000, there was a very slight reduction in payments due to euro devaluation. There was, however, a 10 per cent incentive for farmers who owned under 20 ha, who were paid €165/ha. Both the training course and the planting of trees became compulsory, the inspection rate was halved and penalties were increased.

Teagasc conduct regular reviews of the performance of REPS, comparing REPS farms to non-REPS extensive farms and non-REPS intensive farms, which are defined by the amount of nitrogen they use.[49] The following table shows income statistics for these three groups.

Table 4: Family Farm Income (€/ha), REPS and non-REPS

Year	REPS	Non-REPS extensive	Non-REPS intensive
1999	344	276	702
2000	408	385	797
2001	413	377	947

This table shows that income has increased for all three groups, but the difference caused by REPS is decreasing proportionately. A REPS extensive farmer could earn an extra €68/ha in 1999, compared to only €36/ha in 2001. The income of both of these is much lower than that of intensive farmers, as has been shown earlier in this book. The income per ha of this latter group has grown by a huge amount, due no doubt to their increasingly ruthless productivist farming methods and their state support.

This review also found that REPS farms had a higher stocking rate than non-REPS extensive farmers. My own research supported this statistic. It was only when some farmers' REPS plans were drawn up that they realized just how understocked they actually were, and hence they bought in more cattle. In terms of nitrogen use, REPS farmers used 68 kg/ha, non-REPS extensive used 90 kg/ha, while non-REPS intensive topped the poll at 222 kg/ha. Regarding farmyard investment, REPS farmers spent €47/ha, non-REPS extensive spent €27/ha and non-REPS intensive spent a whopping €87/ha. Investment in machinery had also increased by 36 per cent among the intensive farmers. Overall, REPS makes very little difference to income and environmental behaviour among extensive farmers, but intensive farmers continue to produce more and more all the time.

It was hoped that REPS III could address this issue, bringing more intensive farmers under the rubric of REPS. At this time, REPS has to be compliant with the all-important Nitrates Directive, which will be discussed in depth in chapter four. Negotiations on REPS III were significantly delayed by problems relating to this. The Irish government has sought an exception to this directive, seeking to allow farmers who produce 170–210 kg/ha of nitrogen (N) into the scheme. It is unlikely to be allowed, however. So far, this issue has had to be 'parked', because the EU Commission were not open to it. REPS payments are to increase to:

€200/ha for the first 20 ha
€175/ha for the next 20 ha
€70/ha for the next 15 ha

The maximum payment one can receive is €8550 on 55 ha. The budget for REPS III was €260m for 2004, which was a 30 per cent increase on REPS II. A higher payment exists for those of under 20 ha, which shows that the DAF recognized some problems with the scheme to date and acted upon them. The number of inspections and the rate of penalties have also been reduced. This move has been interpreted by some as meddling by the farming lobby and reducing its credibility as a bona fide programme.[50] The only substantial change has been with regard to supplementary measures. There are now seventeen of these, divided into four categories, and the farmer must choose a combination of them. Without spelling them all out, they are much more in line with environmental desires than previously, including coppicing of hedgerows, installing bird boxes and creating new wildlife habitats.

REPS Uptake

The number of farms in the country now turned over to REPS stood at over 36,000, as of September 2003. This means that about one-third of Irish farmers joined REPS. Take-up was inevitably quite slow in 1994, its first year, but it began to take off in 1995. It was promised that funding for such measures would increase by 40 per cent in 2000–6. This has indeed transpired, but not to the extent promised. The DAF has now revised down its initial uptake projections. It hoped at first that 75,000 farmers would join up, but it appears that they thought this unrealistic in retrospect. They now hope for 59,000 by 2006. Whether or not this is tenable remains to be seen.

Table 5: Uptake of REPS by County (approved plans)
September 2003

County	No. of Plans	No. of Ha
Carlow	336	12,378
Cavan	1336	37,086
Clare	1664	61,859
Cork	2779	108,362
Donegal	3023	112,834
Dublin	83	2881
Galway	4335	140,408
Kerry	2067	112,426
Kildare	435	15,047
Kilkenny	921	39,261
Laois	804	28,079
Leitrim	1189	35,150
Limerick	1068	36,686
Longford	916	27,786
Louth	231	6772
Mayo	4553	147,864
Meath	762	23,367
Monaghan	974	24,131
Offaly	952	33,963
Roscommon	2033	57,790
Sligo	1282	40,213
Tipperary(N)	865	33,370
Tipperary(S)	837	32,772
Waterford	662	29,696
Westmeath	1038	35,146
Wexford	797	30,243
Wicklow	462	20,324
Total	36,305	1,285,894

Source: DAF statistics, 2003.

The uptake has not been evenly distributed throughout the country, as can be seen from Table 5, which gives the latest county breakdown available.

This creates a clear spatial pattern, with REPS farmers being visible predominantly in the western half of the country. The counties of the western and south-western seaboard – Mayo, Galway, Donegal, Cork, Kerry and Clare and the poor border region – Roscommon, Cavan, Leitrim and Sligo, are the ones that have opted for the most transfer of farms to REPS, which generally are the counties with many small farms that are non-dairying, and specialize in livestock and sheep.

The county with the largest number of plans and also the largest amount of land under the scheme is Mayo, with 4553 farms and 147,864 ha respectively. Galway comes second, with 4335 farms and 140,408 ha. Donegal comes in third, with 3023 farms and 112,834 ha. This clearly shows that the areas of the country dominated by profitable farming, especially the south-east, have little enough interest in the scheme. The bottom five are Dublin, Louth, Carlow, Kildare and Wicklow. More money can be made by intensifying rather than the opposite, implied by joining REPS.

The total uptake of REPS at the end of its first phase was almost 45,000, but as can be seen from Table 5, it dropped off on transition to the second round, with a total take-up of *c.*36,000. In a review of REPS conducted by Teagasc,[51] some explanations for non-uptake of REPS II by farmers who had been in REPS I were the penalties they had incurred, the interference with farming flexibility, the backlog of applications and resulting delays, continuing inflation, low farm size makes joining unattractive and poorer farmers were often older and single. Among dairy farmers, high nitrogen use was a problem. Only 18 per cent of all dairy farmers were in REPS. This review recommended examining front-loading, or paying a higher amount to those with smaller acreages. It was identified that there was a need to work on limiting the REPS drop-out rate. One REPS planner found that the rate of payment was too low, the bureaucracy involved too onerous, so eventually some decided that it was 'too much hassle'.[52] It has been reported that there are major delays in REPS payments, especially in the western counties.[53] It has also emerged that over one third of the complaints made to the DAF office concerned REPS payments. While REPS is no doubt a welcome bonus for many small and medium-sized farmers, more radical reforms were necessary to address the deeper structural problems within the CAP, which could no longer be ignored.

More Radical Change in the Twenty-First Century: Decoupling

As already said, one of the major problems in this area is that larger intensive farmers seem to be able to escape any form of environmental controls. The fundamental principles of productivist farming have thus far remained untouched. This era is coming to an end, however. As of January 2005, virtually all of the payments received by farmers have been replaced by a Single Farm Payment (SFP). This payment is promised to last until 2013. This is as a result of the continuing trend of the separation of EU support from actual amount of agricultural goods produced, in other words, the complete decoupling of farm subsidies from production. During these EU negotiations, individual countries had the option of choosing partial decoupling, referring to only certain products, or complete decoupling on all production. The Irish Minister chose the latter option, while some other EU countries chose the former. The same amount of money will go to farmers, but it is claimed that it will cut down on the amount of bureaucracy they heretofore had to endure. It will replace area aid, extensification aid and cattle premia, among others. This will cut down on the amount of applications processed by the DAF from *c.*480,000 to *c.*136,000. The payments for REPS, early retirement, disadvantaged areas and forestry will remain in place. The amount that the SFP will be is determined by the amount received by farmers during the 'reference years' of 2000, 2001 and 2002. In practice, what this means for the farmer is that s/he now receives a set sum regard-

less of what s/he does with the farm. The average payment is
c.€280/ha. This policy move will have been welcomed by
some. One economist said that the amount paid to agricultural
support by taxpayers in Ireland and throughout Europe could
no longer be justified:

> Irish taxpayers and consumers currently pay around £900 million
> [€1143m] annually to farmers through national co-financing of
> EU schemes, through higher food prices and through the CAP's
> share of the Irish contribution to the EU budget.[54]

Matthews goes on to say that 'the farming industry has
become a gigantic social welfare client'. He compares Ireland
to New Zealand, where all agricultural subsidies were removed
in the eighties, and farmers have been forced to adapt by diversifying their activities. In upbeat mood, Matt Dempsey, the
influential editor of the *Farmer's Journal*, said:

> The removal at a stroke of the need to produce while maintaining
> the level of historic direct payments gives farmers for the first time
> real choices in the direction they take their farms.[55]

Despite this approval by agricultural experts, one must remember that this move is accompanied by cuts in prices for commodities such as milk and cereals, which will reduce farmers'
profits. The payment will also be tied to codes of good farm
practice and 1 per cent of all farms will be inspected each year
to ensure compliance. This extensive list of measures makes sure
that farmers obey EU Directives on water pollution, habitats,
wildlife and animal welfare, among many others. Environmental issues loom large, and penalties will be in place for any farmers who flout the rules, most especially the aforementioned
Nitrates Directive. Minister for Agriculture, Joe Walsh, very
much stuck to the official EU line on this during his address to
the Agricultural Science Association annual conference in early
September 2004. He stressed that supplying quality food was no
longer enough from farmers, that society 'has certain expectations in terms of landscape, biodiversity, recreation and leisure',
stressing also that the rules had been in place since 1992.[56]
Meeting these 'expectations' will be a major challenge for many

farmers, especially intensive non-REPS farmers who are not used to environmental regulations. Managing their finances will also be difficult. Assuming they get their SFP cheque with no cross-compliance problems, they will be relying upon one annual payment for a large part of their income, so balancing the books each month could be very hard indeed, especially for those with low incomes and/or large debt repayments.

It is another case of the goalposts being changed for farmers, like the introduction of quotas in the eighties. Some farmers will lose out in this new system, like those who did not claim EU premia for calves and weanlings during the reference years. These farmers may have been intimidated by the paperwork involved or may have fallen outside the information 'loop'. Such farmers, those with low cultural capital, may now lose out on their SFP as a result, unless they have a very good excuse, like illness or a natural disaster. It will be viewed by the most innovative farmers as a window of opportunity to think about diversifying their activities, in preparation for an eventual future without any subsidies at all. It is an opportunity to shed their least profitable activities and focus on their most lucrative ones. It is in their long-term interest to try to upgrade their farming system and produce the kind of goods consumers want, rather than viewing it as a welfare payment to soften the blow of their eventual demise as farmers. After this SFP has been in place for a few years, it will make a fascinating research topic as to, firstly, how the SFP is actually implemented by the state in practice, and secondly, whether or not farmers changed their attitudes and practices in any way after its advent.

Conclusion

Since the foundation of the CAP, farmers have had very little option but to follow the dominant tide of opinion. Decisions are made very far from their reach. At its inception, the CAP was motivated by the ideology of productivism, because of concern at Commission level about food self-sufficiency. This led to the commercialization and intensification of European and Irish agriculture since the sixties. This process remained unquestioned until the eighties, when European agriculture had to be rationalized and costs cut. The pricing policy was gobbling up millions, which were badly needed in more proactive sectors, like rural development and women's organizations. The agricultural surpluses produced had become burdens instead of assets, and these burdens were to be borne by the European taxpayer.

The second round of reform in the early nineties sought to continue along broadly similar lines, but to provide more 'carrot' than 'stick' for farmers to cut their production. This was achieved by framing the policy changes in environmental terms and providing 'green' funding to farmers. However, it is argued here that the new environmental regulations do not in themselves constitute a fundamental re-evaluation of productivist agriculture, as is implied by the use of the term 'post-productivism'. Environmental and welfare concerns have been added on to the traditional price policy, rather than actually replacing it. Instead, these politically expedient regulations established a means of zoning farmers into the role of either market-oriented food producers, or as responsible for environmental conservation.

Commercial farmers are continuing to farm as usual and smaller farmers are being paid by the EU and the Irish state, to merely survive rather than expand their enterprises. While agri-environmental policy might appear to constitute a panacea for rural Ireland at first glance, the question remains as to whether such schemes serve as a genuine vehicle for empowerment for the recipients of this aid, or as a mere welfare payment to help ease the pain of their inevitable departure from the land in the near future? The most recent round of reform is a more serious challenge to the powerful farming lobby, where their SFP is decoupled from production of the same old products and is dependent upon sticking to strict environmental controls. Just how this will transpire remains to be seen.

chapter 4

Green Capitalism

One of the most difficult issues faced by EU policy-makers has been how to reconcile farming with environmental protection. Decisions are made on this by a small cadre of policy-makers and those farmers who are not active in farming organizations are virtually unrepresented in the negotiations that take place. However, farming can pollute the landscape in a number of ways. Food production can be a messy business. The average consumer doesn't think much about this, however, when buying vacuum-packed chicken fillets. The dominant tide of political opinion at EU level now deems it totally unacceptable to subsidize a group who are careless with the environment. The idea of sustainable development (SD) has been wholeheartedly adopted in this sphere and this concept is addressed in the first section of this chapter.

The second section examines REPS in this light. REPS is part of an attempt to incorporate environmentalism into EU farm policy and simultaneously to provide support for those farmers who will be marginalized by continuing economic deregulation. It questions the scheme's effectiveness with regard to these goals. The third section examines the issue of the pollution caused by larger farmers who are outside the remit of REPS. It highlights the current struggles surrounding the Nitrates Directive and how it is being resisted in Ireland. Overall, this chapter is an examination of the 'greening' of capitalist agriculture in Ireland at the beginning of the twenty-first century.

REPS and Sustainable Development

The agri-environmental schemes (such as REPS) that were derived from EU Reg.2078/92 are attempts by the European Commission to solve environmental problems by the application of stricter planning and managerialist methods to agriculture. They seek to prevent environmental problems before they occur and without alienating the powerful actors behind capitalist structures. This process is known as 'green capitalism', because it 'is centred on the avoidance of pollution costs through redesign of processes, rather than on a fundamental reassessment of the need for the products of these processes'.[1] This strategy was extended from industry into agriculture in the nineties. The old approach of cleaning up the environment *post hoc* was becoming extremely costly and bothersome to industry, so it was now in the interests of capitalist social actors to become more 'green'. This instrumentalist approach to environmental management claims centre-stage in debates over land use, leading to the marginalization of the more radical versions of environmentalism. This is now apparently the most expedient approach to solving the problems caused by intensive cultivation of the land.

The manner in which green capitalism provides a common ground between environmentalists and those who hold economic and political power may be interpreted either positively or negatively. On the positive side, it may be interpreted as the green movement maturing and entering centres of decision-making, thereby producing the new consensus of ecological correctness. There is a great danger of political co-optation of the most radical ideas, thereby neutralizing and sanitizing what began as an oppositional movement. Nothing can silence oppositional voices more than claims from dominant political and economic actors that their concerns have already been incorporated into mainstream policy. Some radical environmental groups have been tempted into this managerial approach, because it can get things done, and is a way to attract the public more easily to their cause.[2]

Box 4: Sustainable Development

Sustainable development is the product of the Brundtland Commission's famous report (1987), defined as 'development which meets the needs of the present without compromising the ability of future generations to meet their own needs'.[3] This was the product of a social democratic political philosophy. The environmental havoc caused by hyper-production was recognized for the first time and this was an attempt to come up with a conciliatory solution.[4] The crucial question was how to combine the protection of the environment with economic growth and improvement in human welfare.

If there is one term that has become the keyword of green capitalism, it is SD. It is the policy panacea that purports to incorporate all concerns and render all actors environmentalist. This compromise concept of sustainability emerged in the mid-eighties at a time when the implications of the US Public Law 480 on food dumping and the growing EC food mountains were being exposed.

SD has become such a catch-all phrase that it can apparently be used unproblematically by the World Bank, politically radical community groups and the Irish DAF alike:

> Like 'motherhood' and 'God', Sustainable Development is invoked by different groups of people in support of various projects and goals, both abstract and concrete.[5]

It assumes that science and planning can solve any problems that emerge, with their 'rather benevolently paternalistic imprint'.[6] It also assumes that a certain amount of waste and pollution is acceptable, as long as there are not *too* many Chernobyls or Bhopals. This managed to become the dominant ideology in development circles from the late eighties onwards mainly because it poses no real radical challenge to the most powerful actors in the capitalist system. Rather, it just expresses a concern with the control and management of its worst excesses. It is argued that it is ultimately a product of and a buttress to capitalism, that it:

> purports to reconcile two old enemies, economic growth and the preservation of the environment, without significant adjustments to the market system ... In the sustainable development discourse, nature is reinvented as environment so that capital, not nature and culture, may be sustained.[7]

A more sensible approach is to consider the cultural practices and needs of the real people on the ground who may or may not implement sustainable practices. If, for example, a group of poor farmers depend on cutting down trees in a particular area, or increasing the number of sheep on an overgrazed hillside, the only option for the state is to enforce SD in an authoritarian manner in order to persuade the farmers to comply, because survival issues understandably come before all else for the marginalized in rural communities. The first step towards solving an environmental problem is not to bring in the external experts, as is often done, but to consult with local actors as to what they feel they need to secure a sustainable future.

The Irish government has followed the lead of the European Commission in wholeheartedly embracing this concept. A report was published by the Department of the Environment in 1997 entitled *Sustainable Development: A Strategy for Ireland* which bears testament to the Irish approval of the adoption of the recommendations of the Agenda 21 and Rio Agreements, which followed the Rio Summit in 1992. They declare in this report that they aim towards an integrated approach to the implementation of the SD ideal. In the agricultural arena, the primary stated aim is that of cross-compliance, the idea that no EU aid should be received in member states that damages the environment. The report gives the impression that they have progressed further with this objective than is actually the case. It is only in 2005, eight years later, that cross-compliance was being taken seriously in Ireland, and its real effects remain to be seen. The confluence of agricultural, environmental and tourism policies is very much in evidence in this report. It also seems that it is those areas most dependent on tourist income that are targeted, primarily focusing official attention on such issues as overgrazing in the hilly and mountainous regions of the west.

Since the nineties, we have witnessed the integration of environmental policy with other policy arenas such as those of agriculture and industry. This policy objective is expressed by the EU Commission as follows:

> the targeted sectors would need to start to think of the environment in terms of it being an economic asset as opposed to a restraint to growth. This would challenge farmers and agricultural policy officials to take a longer term view of the viability of the farming sector. Similarly, it would require product manufacturers to build environmental considerations into product design because it makes economic sense to waste less material resources and use less energy.[8]

There is also a stated aim to merge environmental policy with transport policy and tourism policy. As a result of this new approach, many changes have become evident in agriculture. Farmers now must perform varied duties, which are outlined in one EU report:

> farmers have an increasingly important role to play with regard to the structure and use of the land and the preservation and promotion of cultural and environmental assets against a background of growing integration between agricultural markets policy and rural development policy[9]

No contradiction is perceived in adopting several strategies at once, and combining them all in a single framework. For example, the principal developmental objectives for Ireland, as related in a 1997 EU Commission report, are:

- improved efficiency of farm production
- promotion of farming in harmony with the environment
- diversification of on-farm production
- development of the non-farm rural sectors
- improved processing and marketing of agricultural produce
- development of the forestry sector
- income maintenance in the Less Favoured Areas[10]

The complications involved in attempting to combine all of these objectives could hardly be underestimated, especially at local level, where different sets of interests are likely to be in direct conflict with one another.

EU policy on rural development is another arena that has been integrated in recent years with environmental policy. A crucial EU conference on rural development took place in Cork in November 1996, which resulted in a landmark report entitled *The Cork Declaration: A Living Countryside.*[11] This conference was concerned with the development of a multi-sectoral approach to rural development, integrating the non-farm sector in rural areas and also the tourism sector. SD is a keyword of these proceedings, in order to 'protect the quality and amenity of Europe's rural landscapes'. However, oddly enough, the continued intensification of European agriculture was not addressed as a relevant issue at this conference. It appears that this topic is presumed to be beyond the remit of a discussion on rural development, despite its profound impact upon the long-term prospects for the social structure of rural areas. Instead of interpreting SD as a complete re-evaluation of capitalism, it is argued here that it has become a superficial panacea for curing all environmental and social ills. When interviewed for this research, one elected member of the Irish Green Party asserted that the way the term was being sold was as a 'buzzword' which meant 'economic growth without guilt', instead of as a longer term view which combined ecology with social justice. It is worth quoting him somewhat at length to show his experience of the use of the term:

> I think a lot of people who are using it don't have the foggiest clue what it means and I think that a lot of the people who are talking about SD, think that sustainability is a concept which means that we should have our cake and eat it, which isn't the case at all. It means very hard decisions as to how to structure our agriculture, our industries and our lifestyles that mightn't be politically popular but that need to be done to bring about the type of society that we believe is necessary.

Green capitalism purports to simultaneously satisfy all actors within agriculture, but it is incongruous to seek to appeal to rural-regional populism on the one hand, and to industrial-capitalist liberalism on the other.[12] At some point, serious disagreements are bound to occur between various rural actors about the best future for rural areas.

Moving from a general discussion of the synthesizing of environmental policy with other policy arenas such as agriculture, tourism and rural development, we will now turn to our more specific focus: REPS. A new sensibility is being propagated in REPS that obliges farmers to embrace ecology in return for financial assistance. Ecology is defined by Eder as

> a way of looking at nature in which scientific expertise, ethical concerns and aesthetic judgements have been integrated into a coherent ideological framework that provides a common ground for collective actors.[13]

The fundamental primary aim is to combine capitalist agriculture with environmental protection. The landscape itself has become a commodity, primarily to satisfy the needs of the tourist industry. It is no accident that most of the uptake of REPS has been in the west of Ireland, which is scenically attractive and heavily dependent on tourism. Tourism has been well-established in this region since the sixties, and it is hoped by policy-makers that the preservation of ditches, wild vegetation and old farm buildings will increase the value of the countryside for the tourist consumer. It would appear that the aim is to have a win-win situation, where the environment is improved, farmers gain extra income and more tourists are attracted to marginal areas.

The part of REPS that enforces the protection of sites of historical interest and more extensive use of the land is conservationist in style, with tourist revenue as the main financial incentive. This is a very instrumentalist rationale and is aeons away from radical environmentalist thought, which values each part of the ecosystem equally animal, vegetable and mineral.

EU Reg. 2078/92 has very specific aims, but these changed when imported into Ireland, having been interpreted by senior civil servants in the Irish DAF (whose dominant ethos had been modernization). Since the late fifties, efforts had been made to intensify agricultural production at all costs. Farmers had been advised to maximize their yields, increase their herd size, use lots of fertilizer (nitrogen, phosphorus, potassium), knock ditches, enlarge fields, knock old farm buildings and build new concrete ones. The main aim was to increase profits, whatever the consequences. This modernization project was taken for granted for decades so it was quite a change when farms were forced to fulfil environmental protection objectives by the EU in 1992.

The job of civil servants is to create policy using the building blocks derived from their political negotiations. They have to pinpoint precisely what is possible based on the rules and regulations, rather than what they consider to be ideal. EU Reg. 2078/92 was moderated considerably in the Irish case, with organic farming, for example, being removed from the main body of wording. It was relegated instead to constituting one of the additional measures, an elective rather than a core constituent. The aims of REPS as quoted in a pamphlet produced by the DAF are:

- to establish farming practices and controlled production methods which reflect the increasing concern for conservation, landscape protection and wider environmental problems
- to protect wildlife habitats and endangered species of flora and fauna
- to produce quality food in an extensive and environmentally friendly manner.[14]

The DAF staff found themselves in a contradictory situation, being forced to adhere to the new standards of green capitalism. This dilemma is summed up in the following insight from Hannigan:

> As environmental protection has emerged as a significant item on the policy agendas of governments, the state must increasingly balance its dual role as a facilitator of capital accumulation and economic growth and its role as environmental regulator and champion.[15]

The difficulties in trying to balance these dual roles led to the adoption of the managerialist strategy. The Irish civil servants in the Department of Agriculture translated the EU aims into a form that would be more politically acceptable in this country, using the political catch-phrase of sustainability.

The cultural capital derived from adherence to the standards of agricultural science, which is deemed to be inherently superior to that of the farmer, is very important in lending weight or status to the REPS initiative. This emerged when a DAF official was asked to compare REPS to other EU schemes:

> we have all agricultural graduates as planners and all the plans are submitted through them. I think in some countries in the EU, the farmer submits his own plan (laughs). There's no way in the world, with the type of scheme that we have, 'tis complicated enough, they couldn't do it. We're way ahead of the others.

Therefore the intuitive ways that farmers have of interacting with the environment are dismissed as folly. However farmers may have other motivations besides profit maximization, or may interact with their land and the overall environment in different ways than that dictated by economistic thinking. Local knowledge systems are generally ignored or dismissed as idiosyncracies, as little more than obstacles to progress. Little value is placed on people's everyday perceptions and non-expert experience, or on the realm of 'intuited nature',[16] which has to do with integrity and emotions. This approach could also be said to result from the dominance of a male-biased scientific establishment wherein little or no value is placed on that which cannot be measured.

REPS is based on the results of soil samples, the calculation of stock density rates per ha and amount of organic nitrogen produced per ha, producing a type of knowledge that is centralized and officialized. In the administration of REPS, science has been allocated an enhanced role and a new layer of expertise has been institutionalized in the structure of Irish farming. REPS plans must be drawn up by an approved planner, the main one being Teagasc, as well as many private planners. These scientific experts 'search for quantified indices of phenomena, rather than looking for the socio-cultural context and practices of the actors' actions that surround and shape underdevelopment'.[17] However, the officialized scientific terms used are very foreign to most of the farmers interviewed during this research.

An agricultural science education has become almost compulsory in order to be viewed as a good farmer. This official language controls the social agenda by

> laying down the dividing line between the thinkable and the unthinkable, thereby contributing towards the maintenance of the symbolic order from which it draws its authority.[18]

Science thus becomes a self-fulfilling prophecy in this field, where one is progressive only if one adopts scientific methods. The end result is that it serves as a gate-keeping mechanism that includes the progressive and excludes the traditional farmer.

The ideas about the environment that drive official debates are very much social constructions. The way that environmental issues are understood is 'as much or more a matter of the social construction and politics of knowledge production as it is a straightforward reflection of biophysical reality'.[19] Ecological problems are organized around 'emblems',[20] or issues of great symbolic importance at certain times in certain places. Some of these have been soil erosion in the US in the dust-bowl days of the thirties, nuclear power in the seventies in Europe, and acid rain (especially in Germany) in the early eighties. The particular emblems that have served as symbolic motivators in contemporary Irish agri-environmental policy are water pollution, especially of the fishing lakes in the west, and the overgrazing of the hillsides in the commonage areas of the west.

There has been an increasing amount of concern among agricultural and environmental experts about deteriorating water quality and the farmer's role in causing fishkills in the nation's waterways. One DAF official said:

> Water quality was a major issue, especially on the good fishing lakes in the west of Ireland, which are exceptionally pollution-free by any standards in Europe, but nevertheless they were deteriorating. The salmonid waters in Ireland are exceptionally good and we wanted a high uptake in the West of Ireland.

REPS was an opportunity, therefore, to take action against the eutrophication, or over-enrichment with nutrients like nitrates and phosphates, of lakes and rivers. A Nutrient Management Plan was drawn up for REPS farmers that attempt to ensure that their farming practices produce no more nutrients than can be absorbed by their land, i.e. that they do not pollute their environment. However, the main uptake of REPS has been among smaller, non-dairy farmers, who produce very little pollutants. It is not at all clear that REPS is actually helping to improve water quality, as it is claimed by the DAF that research does not exist on the matter. When asked if any information was available on water quality in areas of intensive dairying, this Department official diverted the discussion to research that is currently being carried out in probably the busiest tourist site in the country:

> in the Lakes of Killarney, there's a big project there in conjunction with the Lough Laune Environmental Group. There's £700,000 [€888,800] of EU money tied up in a project ... to try and clean up Lough Laune.

Clearly, then, areas of high tourist activity are prioritized for attention. This indicates the double standards that exist, with the large dairy producers, who produce the most pollutants, subject to little or no environmental controls up until 2005. Fourteen years have elapsed since the agreement of the Nitrates Directive, which will be discussed in much more detail later in this chapter.

The second emblem is overgrazing by large numbers of sheep on commonage lands in the west. The generous headage payments paid by the EU for sheep and cattle were the main reason for the high stocking levels, so agricultural policy-makers are attempting to solve a problem that is ultimately of their own making. Over the past few years some incentives were introduced to cut down on stock numbers but, according to a DAF official:

> it was quite clear to everyone that we weren't solving the problem. We had a number of EU audits as well looking at the issue on commonages and there were hundreds of complaints coming from environmental organisations and individuals to our Dept and to Europe.

The Irish DAF has been under pressure to produce results in order that EU co-financing will continue. These two main emblems helped to shape REPS as it now exists.

This new concern about protection of the environment dove-tailed in Ireland with the concern for creating alternative income for marginal farmers. It seemed the perfect parallel solution to the two problems, meeting the needs both of the EU and the State and providing funding for farmers to co-operate. The DAF there-fore attempted to introduce an environmental aspect to Irish farming, combined with the provision of social assistance to smaller farmers. This latter element was enhanced in the Irish case. Environmentalist measures became a way of framing EU aid, which could be read as an attempt to ensure social content-ment and harmony and moderate the excesses of capitalist accu-mulation. According to one Teagasc REPS planner:

> It's more a social payment than an environmental payment. ... Money will always be pumped into the countryside, to keep the social fabric intact.

It is suggested here, therefore, that it is part of a legitimiz-ing strategy which maintains the deepening inequality of capi-talist agriculture. Teagasc in general have give REPS their seal of approval. A Teagasc REPS advisor claimed that REPS is regarded as the most successful adaptation of EU Reg.2078/92 in Europe, with agronomic, socio-economic and environmental benefits for the country. They have strongly recommended REPS to drystock and suckler farmers, arguing that it con-tributes to income generation and maintaining population in rural areas. They usually frame that approval in terms of effi-ciency, rather than meeting any ecological objectives *per se*. In fact, their attitude to conservationists can be seen to be rather hostile. A senior farm advisor from Teagasc stated:

> I was at a conference on the environment recently where fellows with beards and English accents were giving out about farmers farming the land intensively. They don't seem to understand that people have to make a living.

In a similar vein, a planner in Skibbereen said:

> At the start they [farmers] thought it was 'back to the future', that they were going backwards, but now a lot of the farmers look on it as an efficient system ... It takes disaster to make people see how they should keep their margins tighter, how efficiency should come into a

farming system, but definitely there was a lot of luxury spreading of fertiliser, which was very inefficient and they see that now.

The categories of progressive and non-progressive are very firmly rooted in Teagasc thinking. When asked to describe the profile of a typical REPS farmer, this was the response of one planner:

> Mostly they haven't progressed much since about the mid-sixties. They're still stuck in that era. Most of the progressive farmers have done their slatted sheds, big buildings, increased their stock and reclaimed land, these fellas haven't. Either they didn't have the money at the time or the inclination or whatever.

Said another when asked the same question:

> They're not what we call viability farms, they're not commercial farms really. They're not at the cutting edge.

With the input of all these sets of actors, a policy initiative like REPS can be transformed by percolating down through the various levels from the EU to the farmer.

It would be inaccurate to claim that a completely new era of policy-making has emerged since the nineties, but rather that the ethos of productivism has just adapted to changing times and economic pressures within agriculture by adopting an uncontentious version of environmentalism. This approach is consistent with continued productivism because it does not challenge intensive land use, or the concentration and specialization of production. An advisory role has been created for agricultural science experts and they retain their primacy over farmers' perspectives on land use. Intensive farming and environmental protection are encouraged concurrently in different zones. There is a major contradiction between the encouragement of intensification on the one hand and the revaluing of peasant farming on the other. This contradiction is managed by the fact that agri-environmental schemes apply to those at the non-productive end of the scale of farming while up to now, productivist agriculture continued unabated. The EU budget was no longer to be spent on subsidizing overproduction by large farmers, but instead on producing environmental goods for the consumption of the increasingly disgruntled European urban-dweller. The Greens were seen to be getting things done in terms of inducing farmers to adhere to ecological parameters, at a time when it was politically popular to get tough with this sector. This is especially true in Ireland where there exists a strong urban-rural divide.

We can also witness a similar zoning occurring on an international scale. Our environmental concerns may easily become ethnocentric and global inequalities cannot be omitted from the equation. While agricultural productivism is under ideological attack in Europe and the US, we should not forget that it can be exported elsewhere:

> There are sufficient countries very keen to attract overseas investment to allow the exploitation of land and labour to meet the food needs of advanced economies.[21]

It is thus the existence of poverty in other parts of the world that allows such arguably esoteric issues as environmental protection to be prioritized as a political issue in richer areas such as Western Europe.

Social inequality is also reflected in changing food tastes. Goodman and Watts argue that agricultural restructuring produces opportunities for both continuity, with the continued mass production of cheap standardized food for the mass market, and change, with the development of a multiplicity of newer niche markets, e.g. for convenience foods and organic food.[22] While it is important to be aware of issues to do with international rural restructuring, let us not forget that 'external forces are mediated at the level of the household and that structural change in agriculture is also conditioned by the attitudes and behaviour of farmers as individuals'.[23] It is to the everyday experiences of these individuals that we now turn.

REPS Farmers and Pollution Control

According to a senior civil servant who was involved in devising REPS, the aims of REPS are, in order of priority: environmental, to solve farm pollution problems; agricultural, to create a more extensive agriculture; and socio-economic, to distribute income to those most in need. REPS aims, therefore, to discourage farmers from producing pollutants that damage soil and water, and from producing more of the agricultural commodities that are currently in surplus, primarily beef and milk. It also aims to compensate with direct payments those smaller farmers, especially in disadvantaged areas, who fare badly from the EU pricing mechanism. The opinions of the farmers themselves are given on these aims, in order to compare and contrast the original objectives of the scheme with its everyday implementation.

From the Irish farmer's perspective, the core environmental issue is balancing the numbers of cows and cattle against the nutritional needs of the soil. The more stock a farmer has, the more need for a lot of grass to provide winter fodder for them, therefore the more likelihood of using a lot of fertilizer to seek to augment grass growth, and the more likelihood of causing leaching of phosphates and nitrates into groundwater. Also, the more stock, the more manure produced, the more potential for slurry runoff, and the more need for sophisticated slurry disposal systems. Therefore, the issues of amounts of pollution produced and agricultural production are difficult to separate. A basic set of questions posed to the farmers therefore sought to find out what their practices were in this regard before and after joining REPS.

It emerged from this research that the kind of farmer who is in this scheme, according to a senior farm advisor:

> isn't intensive, so he's probably not polluting anyway ... this doesn't really reach the intensive farmer, those that are using lots of slurry and fertilizer.

Their stocking levels are relatively low, so their herds do not produce much slurry and they are not under massive pressure to generate high grass yields, as the intensive farmer would be, so they generally are not using excessive fertilizer. A comparison of REPS and extensive non-REPS farmers conducted by Teagasc found that the same amount of organic nitrogen and phosphorus were produced by both groups and there was only a minor difference in chemical nitrogen rates.[24] However, this research found that the inheritors did improve their farm management methods, like accounting and record-keeping, but most already operated pollution control facilities because this is part of good farming practice. It appears that most of the farmers in the study group already had a well-organized farmyard, with little or no slurry runoff. Those who did not used the REPS money to help install pollution control systems, which several of them predicted would be required of all beef and milk producers in the future. They did feel that it encouraged them to manage their farming a little better, reduced their nitrogen costs, and encouraged them to get better use from slurry, organic fertilizer. They would not have bothered joining REPS if they had no system in place. Some did install a slatted house for cattle, costing €40–50,000, which is apparently *de rigueur* these days. This is a very major expense for farmers of this size and type. One farmer explained:

> I built a slatted house ... 'Twas part of the farm plan but I couldn't afford to do it until I was guaranteed the REPS that May, that pushed it on for me. You see, I have three boys and three girls and I couldn't afford it.

They all said that they would have done this anyway, but REPS helped to pay for it. One young man reported:

> My father was farming before me and things were run down, so it all had to be done. So REPS helped to pay for all that

So it was just completed sooner than it might have been otherwise. These farmers were using REPS like any other grant, to pay back bank debts that had been incurred to upgrade their farms. However, most reported that they had only to make minor changes, like perhaps concreting their farmyards, and they continue to farm as they have always done. There was evidence among these farmers of an informal code of good practice in this regard that appeared to them to constitute 'just common sense', which was an oft-quoted phrase. They never spread much fertilizer because they could not afford to spend much money, so it made economic more than ecological sense to be parsimonious in this regard.

Some of the inheritors, however, did think in ecological terms, as can be seen in the following examples:

> I can't accept this system of set-aside and all that. I can't see it making sense to pump nitrogen and sprays into one field and let the next field idle. Surely it would be better for the environment if half the fertiliser was put out over the whole thing, with half the chemicals, and take half the tonnage off it.

> I totally disagree with spraying around the headlands, because we have a small child there and one day we want her to see the ecology of the area as well, the flies and the butterflies ... If you put Gramoxone or Roundup on the dykes, you're killing everything that's there, so I think it's totally wrong.

If the inheritors' actions are mainly guided by economic necessity, those of the buy-ins are usually motivated by high levels of ecological awareness rather than financial considerations. This group are almost all from an urban background, middle-class and university-educated. Several, especially the Germans, earned their main income from business interests back in their place of origin. Nearly all of them have signed up to the additional measure that pays extra to get into organic farming and many of them think that REPS does not go far enough. A German couple sent a letter to their REPS planner telling him why the burning of the hillsides should be completely banned, rather than just restricted. Their high level of education also appears to prevent some from having to comply with certain measures. An English woman told me how she was

> allowed a dispensation not to put lime on the meadows because I quoted all the Latin names of the flowers that were growing there that would be killed by lime and he let me away with it. I don't want to have to go around checking the pH of the soil on a bog!

The standards they kept were higher than those required by REPS. One German man who raises organic lamb uses German standards of animal welfare, keeping all animals on comfortable straw bedding. Both felt that the planner was only interested in implementing REPS regulations, even when their alternative practices were superior. Another English woman, an An

Taisce activist, felt that they were 'getting money for old rope'. This group especially had many criticisms of the scheme, and also suggestions for improvement.

It appears to be rather nonsensical that the smaller extensive farmer has to bear the brunt of environmental controls under REPS. The farmers interviewed here also recognize this. Several commented that surely the scheme was more or less meaning-less if everybody didn't have to take measures for pollution control. The positive effect from it can only be at best patchy, at worst negligible. Some farmers argued:

> If more people went into it, it would be better for the environment. We're the only ones in this whole locality that are in REPS and it isn't enough. For the little bit of good that we do, if all around us aren't equally doing it, it would be more effective with more in it.

> Well, one person is doing it and fellow next door can do what he likes. If everyone isn't in it, how is it going to work?

> A fault I find with it is that everyone should be joined it. I'm mak-ing an effort here to make the place reasonably tidy and the guys down the road ...

If REPS does not reach those major polluters, who or what does? The managerialist approach of the Irish state relies on legal control measures to punish polluters, after the harm is done. This 'after-the-event' approach treats any misde-meanours as exceptional mistakes which are to be dealt with on a case-by-case basis. It perceives a zero-sum trade-off between economic growth and environmental protection.[25] Pri-ority is usually given in Ireland to monitoring pollution in tourist areas such as the Lakes of Killarney. This is a mild, inof-fensive approach, which is by no means strict on polluters. It is 'problem-centred', seeking to solve them in an *ad hoc*, piece-meal fashion. This leaves capitalist production intact and places the environmental problem-solving in the hands of des-ignated experts. This research therefore postulates that REPS is inefficient as an overall pollution control mechanism.

Is it any more effective in cutting back on beef and milk pro-duction and encouraging farmers into more extensive methods? Virtually none of the farmers interviewed for this research cut back on cattle numbers in order to join REPS. The majority are farming the same way they always have been and producing the same amount of commodities. One asserted that he wouldn't have joined if he did have to cut back on stock numbers:

> No, if I had to I wouldn't have joined in the first place. I took extra land to make myself qualify for it – I wouldn't have scaled down stock-wise by any means.

Indeed, ten farmers out of a total of sixty replied that they had leased extra land in order to become more extensive overall. Paradoxically, then, for a policy that hopes to keep more people on the land, instead it has encouraged some farmers to increase their farm size. The existence of REPS perhaps provides an incentive for older farmers to lease their land to their younger, more active farming neighbours. This is almost certainly more lucrative for these older farmers than joining REPS.

Another ten farmers admitted that they actually had more cows and/or cattle now than before joining REPS. Once they had their stocking rate assessed by Teagasc, they realized how understocked they were, so they bought more. For example:

> No, I have more cattle, I've doubled my cows from sixteen to thirty-two and their followers. I wasn't heavily stocked and it helped me to build up my herd.

> No, I'm milking a few more cows actually, because I was entitled to milk a few more. They told me how many I was allowed and I'm well within my limits.

In these cases, the farmers' encounters with the Teagasc advisors led them to think in a more productivist fashion than they previously had. Paradoxically, they were encouraged to interpret REPS in a productivist manner, rather than as a means of raising their environmental and ecological awareness.

REPS was meant to be part of a package of measures, based on EU Reg. 2078/92, to cut back on production and encourage ecological use of natural resources. The same amount of, if not more, cattle and milk was being produced by the majority of farmers interviewed. Overall, extensification could only be said to have occurred if the land was now supporting less cattle than in the past. Considering how marginal the previous owners of the land probably were, this is highly unlikely. Teagasc's own research on the impact of REPS found that the average stocking rate for REPS farmers was 1.26 Livestock Units per hectare (LU/ha), compared to 1.3 for non-REPS extensive farmers, and 2.48 LU/ha for non-REPS intensive farmers.[26]

In general, the opposite trend of intensification in farming is still increasing in Ireland. Average farm size has increased and production is becoming more concentrated. For example, there has been a huge drop in the number of milk producers over recent decades, from 77,000 in 1966 to about 15,000 in 2004. One DAF official recognized the contradiction between the economic and social goals of EU farming policy. He stated:

> There is a problem here because the basic philosophy that seems to be coming out of the EU is to get bigger, more efficient units operating to give family farms a decent standard of living, with bigger and bigger units to produce cheaper and cheaper food, harmonising with world prices. There's a contradiction there in that how do you safeguard the countryside, where expensive infrastructures have been put in place and this policy is taking people away from the countryside, resulting in waste of money because there's no-one to look after it all.

The farmers in this study group were also very concerned about the level of concentration of production that they see occurring around them in their local area. One young man in his twenties, who owns 45 acres and has leased 38, says:

> You have to be big or you can't survive at all. The bigger fellows are eating into the smaller fellow's land. It's nearly impossible now to lease land with quota because of the high prices of renting. Farming is barely viable now.

This is especially the case in the dairying sector. It would seem that REPS has little or no impact upon this trend which exacerbates rural depopulation, because the income they receive from it is not sufficient to ensure their survival on the land.

The rather simplistic linking of ecology with extensification has also been questioned. It cannot be assumed that a reduction of stock numbers is necessarily environmentally beneficial in all areas of the EU.[27] It may well be the case that too *little* use of the land may be the main problem in some areas of the large and ecologically diverse land mass of Europe. This blanket approach has led Symes to conclude:

> There is need for a much closer fit between policy and specific regional circumstances. Off-the-peg solutions applied across the highly differentiated surface of European agriculture have proved increasingly unsuitable.[28]

This view was shared by a local An Taisce environmentalist activist in west Cork:

> You probably would need more localised planning, these rules about spreading slurry on certain days or months, that has to be brought down to local soil situations. Sticking to them might not be possible in terms of weather. There has got to be a more localised application of rules.

The varying local climactic and topographical conditions, geographical and cultural diversity therefore makes it difficult to impose a unilinear development path, which may or may not address the real needs of the target populations. Portela expresses this well:

> Agriculture is spatially grounded and climatically dependent. Being nothing without plants, crops, animals and herds, it has a biological side. The interplay of these traits inevitably feeds regional, micro-regional and local dimensions ... a common policy has to be built from 'below', recognising the specificity of the place, the time and the intrinsic craft basis of agriculture.[29]

When Teagasc officers were asked by this researcher what difficulties farmers themselves had with REPS, they reported that the question of control over their farming practices was a major issue for them:

> Farmers don't like to be dictated to as to what they can do. They were always raising sheep on that bloody mountain and who is this fella to tell us that we can't do it ... Their main reservation was that they were going to be curtailed in the way they could farm. It is the same in the SACs where farmers are told you can't plough that field or you can't put forestry in that field ... I suppose people will eventually have to accept that while they own the place, they're only the guardians of it, they can't really do what they like with it.

This senior farm advisor is indeed very familiar with farmers' attitudes to the land, based on his own personal experience of what Arce terms 'the social dust of the field'.[30] Similar concerns about rights were voiced at a public meeting on REPS in Maam Cross in Connemara in early December 1997. A journalist reported on the proceedings at this event:

> 'We've been given a lot of *dos* and *don'ts*,' said one farmer, 'Are there more? When will it stop?' They complained bitterly of losing ownership of their land and livelihood. A warning that Connemara farmers would not become like 'Red Indians or Aborigines living in a reservation' was cheered. 'Once you have to ask permission you've accepted you have no rights on your land anymore.'[31]

This lack of control over production decisions has been a feature of Irish farmers' experience since first joining the EEC. Because their time-honoured ways of managing their farms are now undermined in the face of growing market competition, they are 'being systematically stripped of their sense of identity and purpose'.[32] This EU control has intensified since CAP reform and the introduction of quotas. The more recent addition of agri-environmental measures are a continuation of this trend, and these have been read as 'a financial trade-off between individual rights and community benefits'.[33] It is difficult to see how priorities can be decided upon, without glossing over the massive differences that exist between regions of this vast space.

The income farmers receive from REPS is however very welcome in most cases. When asked what they considered to be the advantages of REPS, many of the inheritors reported that it was a reliable income in very unreliable times, the 'C.I.P., the cheque in the post'. They were being paid to do 'what they should be doing anyway'. They are glad to be paid to 'clean up their act' and to use slightly improved farm management strategies. Most were using it simply as another source of income, rather than as an environmental measure, *per se:*

> Well, you'd have the money to pay bills. Everything is so dear and farming is gone downhill.

> I knew one fellow who had a bill at the creamery and he was never

able to clear it until he got the REPS cheque. It has certainly helped the low-income farmer.

Twenty-five out of sixty of the interviewees, or 42 per cent of them, had another income coming in to the farm household, earned by some member(s) of the household. This was viewed as being needed to survive. Some worked in construction; others with the County Council. Some wives worked in hotels, as hairdressers, or in the civil service.

> I certainly wouldn't survive on farming alone. I have a job as well. Unless you have a massive amount of land and a lot of cattle, you couldn't survive on farming alone, even with REPS. You'd be living like a monk.

> Well, I heard last night of a man with 100 cows getting out of farming because there's no money in it. There's no money in what I'm doing either, it's just a pastime, walking around in the evening. There's nothing to be made in it.

The REPS income means a lot to some, on drystock farms especially. It is viewed by many as as a good stop-gap measure, making some difference in the short-term, especially in the cattle sector.

> It's a help financially all right, that you have an income for 5 years. Everything helps, its an extra few pounds. When times are slack, you turn to every option to get money.

> REPS is a big part of my income, that and headage and the suckler premium and that. If there were no premiums, I wouldn't survive here. With a bunch of animals you don't know until you have it in the post.

Most felt that it wasn't enough money to prevent farmers leaving the land if they so decided. Their long-term future was more dependent on the quota system and EU pricing mechanisms for commodities.

> REPS isn't going to prevent a fellow from selling, if he thinks his enterprise is too small and that he'd work out better by selling it off ... It's not enough money.

> With REPS, unless there's a more competitive amount of money I don't see how it's going to make any difference at all. The big farmers in the Midlands aren't going to be interested in five grand

It does not reach the poorest farmers and the older farmers, those most in need of help in an era of spiralling problems for this sector. This may be because of the economic costs involved in, for example, converting the farmyard facilities to REPS requirements, or the potentially prohibitive cost of the initial planning. It may require a level of education and know-how that is beyond the realm of possibility for many marginal farmers. This was verified by a Teagasc advisor: 'the older fellows who just don't want to know about it, because it involves too much work'. Despite the fact that many of the practices actively endorsed by REPS were common-sense to the older country people, like spreading lime on the land, they are informally discouraged from joining this scheme and encouraged to take early retirement and lease their land.

Some viewed REPS as being similar to social welfare, the only difference being that one had to perform some duties for this, whereas welfare is for no work at all:

> The dole is money for nothing, but this is attached to the land and work. This is another sort of dole in a sense, but its linked to doing something. It's some way productive.

> Well, it does give people an income in the short term but in the long term I think it will do more harm than good. People will get dependent on easy payments like this, instead of doing something more productive on their farms.

While REPS is not exactly the same as social welfare, it shares several features with it. Those who are encouraged into REPS are those that are marginalized by the processes of agricultural restructuring throughout Europe. It does not, however, reach the older farmers or the smallest farmers, who own under fifty acres.[34] It provides a minimal income for small farmers, firstly to help quell social unrest, and secondly, to ensure that they can continue to serve as consumers of the goods produced by capitalism. Welfare serves the function of exerting social and political control over this group.[35] In the welfare state

> material benefits for the needy are traded for their submissive recognition of the 'moral order' of the society which generates such need.[36]

In REPS, farmers are accountable to the state for the practices that they adopt on their farms, in return for financial reimbursement. Their social role is more tightly circumscribed, controlled and monitored than it would be otherwise. It is also politically expedient to use the rhetoric of sustainability and ecology to mollify the least radical wing of the European Green movement. However it does little to improve small- to medium-sized farmers' survival chances, as the amount of money they receive is quite small compared to the costs involved in complying with the rules. It has little or no impact upon their long-term future, despite the rhetoric that deems it to be ideal for marginal people and places.

Non-REPS Farmers and Pollution Control

Having looked at the REPS farmer, what of the farmer who is not tied into any agri-environmental scheme? We can see that the uptake of REPS in those areas of Ireland with intensive dairying, the east and south-east, is very low compared to the west. Irish farmers who are not in REPS have been subject to little or no environmental regulation. One DAF official admitted that the intensive farmers who are not in the scheme 'are potential minefields as far as pollution is concerned'. Agri-environmental policy uses less than 5 per cent of CAP spending, so its impact is bound to be minimized by broader CAP policy.[37] REPS is not geared toward the intensive farmer, who can earn much more money by farming intensively, and usually in the dairy sector. The price support system in EU policy is their mainstay and schemes such as REPS do not interest them. Considering how much they are currently earning, they would need to earn at least a comparable amount, if not more, from an agri-environmental scheme. Of course, it would also require them to re-think their productivist ethos. The fertilizer restrictions that have to be adhered to in REPS would be anathema to the large dairy farmer, so they generally do not join. This problem was recognized by Teagasc personnel, as can be seen from the following quotes from planners:

The big fellows aren't getting into it at all because they're too intensive and £50 [€63.50] an acre doesn't in any way compensate them for the losses they'd have if they were to get into the existing scheme, they'd need four or five times that ... It would be nice if pollution could be controlled in the more intensive people, because that's where the bulk of the nutrients that go into our water come from.

To improve the water quality in Ireland, they'll have to bring in the medium to large dairy farmers, and the way of bringing them in is maybe to change the nitrogen specifications.

Intensive non-REPS farmers are using 206 kg of chemical nitrogen per ha., compared to 59 kg/ha and 85 kg/ha for the REPS and extensive non-REPS farmers respectively.[38] Encouraging large commercial (especially dairy) farmers to farm in a less environmentally aggressive manner has been a political problem that has been downplayed so far. Large farmers are quite a powerful political group in Ireland and it is inadvisable for any political representative to challenge them directly. Action on farm pollution has now been forced into place by the cross-compliance requirements of European policy, over the heads of Irish politicians. This is implicitly endorsed by the Green Party, as in the following quote from a Green TD:

We'd like to see REPS as the cornerstone of European policy rather than an add-on or a conscience-salving measure.

This more radical type of approach has been termed 'green recoupling', which is 'a system in which all supports to farmers are delivered through environmental schemes'.[39]

However, in total opposition to this environmentalist view, a Teagasc officer criticized REPS for being 'too all-embracing', that farmers would prefer to comply with a few of the REPS measures, while continuing to farm normally without causing pollution. The ideal picture for him would be to have a few core measures and several optional ones. This Teagasc representative obviously retains his productivist mentality, because he wants to see the environmental content of the scheme diluted even further. No doubt environmentalists would interpret this as wanting to have your cake and eat it too. The situation has now emerged where the small and medium farmers who are in REPS are being strictly monitored for pollution and large dairy farmers are hardly subjected to any regulation, except perhaps in the rare case of a clear breach, like a fishkill, and, crucially, where the source of the poisonous emissions can actually be legally proven.

The state environmental regulatory body, the Environmental Protection Agency (EPA), which was only founded in 1992, carried out a major survey between 1995 and 1997 that found Irish waterways to be overall satisfactory in a European context. Although water pollution levels in Ireland are nothing like as serious as in the Netherlands, for example, there has been an increase in slight to moderate pollution in Irish waterways, with several fishkills being probably attributable to farmers. There has been a 10 per cent increase in the amount of water pollution in Ireland, with one-third of all waterways now being termed polluted and a steady decline in unpolluted rivers and lakes. The main culprits are agriculture, septic tanks, detergents and sewage. Another problem is siltation in the west caused by sheep over-grazing and bog and forestry development.[40] A 2004 EPA report found elevated nitrate levels in 23 per cent of all drinking water; animal wastes and faulty septic tanks are blamed as the biggest cause.[41]

It is extremely difficult to apply the 'Polluter Pays Principle' in the case of farming. The EU Commission states:

> While acknowledging that the Polluter Pays Principle is accepted, it is difficult to implement in relation to nitrate pollution. This is due to the difficulty in attributing responsibility for it among farmers, because of the diffuse source of pollution and the complex process by which nitrates reach the groundwater.[42]

The problem lies in legally proving the precise origin of any episode of farm pollution, especially since 'most contaminants released by agriculture are also released from other sources'.[43] When generalized legal penalties (as opposed to individualized ones) are put in place, Marsh argues that 'a general charge may punish the innocent and let the offender off too lightly'.[44] The Department of the Environment in 1996 deemed two-thirds of fishkills to be 'not attributable to any single source', hence no legal action was taken.[45] It is within the power of local author-ities to enact by-laws to regulate agriculture and to enforce Nutrient Management Plans, but at the time of writing only five county councils out of twenty-six had put in place such by-laws: Cork, Cavan, Westmeath, Tipperary North and South Rid-ings. Cork led the field on this, which was as a result of some major incidences of farm-related pollution on the river Lee.

The EPA introduced a strict licensing system in 1997 for 'intensive agriculture', but strangely, this is defined solely as pig and poultry production.[46] Attempts were made throughout the nineties to initiate this system, yet it did not make any progress for several years, 'largely due to the agency's naivety in under-estimating the incisive and well-organised opposition of the farming lobby'.[47] This system, Integrated Pollution Control (IPC) Licencing, actually excludes dairy production to date and focuses on pig and poultry factory farming units in the Cavan/Monaghan region, where the lakes and rivers are noto-riously polluted by excessive phosphates whose source is pig slurry. Only 34 per cent of the water in this region is unpol-luted.[48] Lough Sheelin in Co. Cavan is highly polluted, and its stock of trout has fallen from 115,000 in 1979 to just 7000 in 2002.[49] The dramatic statistics from this region naturally mean that it has to be prioritized, but the exclusion of intensive dairy farming from any regulation by the EPA is inexcusable. Keo-hane, following detailed observation of the outcome of an EPA oral hearing in Limerick, asserts:

> the EPA acts as the choreographer in a power play enacted to deceive those whom it is supposed to serve, and to corrupt one of the fundamental institutions of democracy.[50]

While the number of fishkills has dropped significantly since the eighties, water pollution continues to be a serious problem, with thirty-seven cases occurring in 1999 alone.[51] These are caused by silage effluent, slurry discharges, excessive fertilizer use, industrial and domestic pollution. The single most impor-tant ruling from the EU on this issue is the Nitrates Directive.

The Nitrates Directive: A Political Minefield

Ireland is in big trouble with the EU Commission. The source of the problem is the lack of action taken on the issue of water pollution and, in particular, on implementing the legally binding Council Directive 676/91, or what is commonly termed the Nitrates Directive. According to its own text, its objective is 'reducing water pollution caused by or induced by nitrates from agricultural sources and preventing further such pollution'.[52] It goes on to say:

> common action is needed to control the problem arising from intensive livestock production and agricultural policy must take greater account of environmental policy ... The main cause of pollution from diffuse sources affecting the Community's waters is nitrates from agricultural sources ... it is important to take measures concerning the storage and the application on land of all nitrogen compounds and concerning certain land management practices.

In March 2004 the European Court of Justice found that Ireland was the only EU country to fail miserably in this regard. No nitrate vulnerable zones had been earmarked nor action programme established in early 2004, never mind by 1995, when it was required by the Directive. This problem also caused delays in negotiations over REPS 3. This ultimately means that Ireland now absolutely has to obey EU law, or the Irish taxpayer will incur serious fines, thus bankrolling the intransigence of large Irish farmers.

Apart from joining the EEC itself, and possibly the introduction of quotas, this is probably the single biggest issue to affect Irish farming. From January 2005 farmers receive a Single Farm Payment from the EU. This is dependent upon adherence to a long list of rules, the most important of which derive from the Nitrates Directive. The issue of water pollution by farmers has been floundering at the bottom of the Irish political agenda for many years and it is only now that serious action has been taken. The Nitrates Directive was introduced in 1991, following concerns about the health effects of excessive nitrates in the EC's drinking supply. It legislated that the nitrate content in water should not exceed 50 mg/litre.

This primarily affects intensive dairy farmers. It rules that farmers cannot produce more than 170 kg/ha of nitrogen, which is equivalent to 2 cows/ha. This limit has been the cause of a massive row in Irish agriculture, with government representatives caught between their loyalty to farmers and their need to keep EU rules. They are forced to agree to this limit, but they are seeking a derogation for the Irish case, of 250 kg/ha, or 3 cows/ha. This is unlikely to be successful, as the only precedent is an extremely limited one from Denmark. It is, however, the 'last chance saloon' on this issue for the Irish farming lobby. An added problem is that a decision on the matter could take years, making it difficult for farmers to decide how to cope in the uncertain interim period.

A key issue here is the debate that is occurring regarding the scientific basis of the Directive.[53] Teagasc has just published the results of a recent report on nitrates in Irish water and the connection with farming.[54] They support the case for a 250 kg/ha nitrogen limit for Ireland, given its soil, weather, and land use. Soil in Ireland is often quite wet, which would prevent leaching of nitrogen. Also, our high level of rainfall means that any leaching that did occur would be diluted and washed away quickly. Finally the fact that 90 per cent of Irish land is under grass is an advantage, as they have found grassland to be very effective in preventing nitrogen leaching. There seems to be an implicit criticism of the blanket approach of EU law, which is deemed insensitive to localized environmental conditions. Notwithstanding these results, the input of an Irish research organization that supports intensive agriculture and challenges the basis of the Directive is not likely to have much impact at the level of the EU Commission, which legislates for the entire EU.

The Directive has serious practical implications for farmers. Restrictions are to be imposed on spreading slurry and chemical fertilizer and manure will have to be stored for about twenty weeks of the year. These rules are similar to those tolerated by REPS farmers since 1994. Presumably, the more responsible and environmentally aware farmer will have already adopted practices that are similar to this code of good practice. The big change is that which was previously recommended is now legally mandatory, and their living actually depends on it. This new edge is very worrying for farmers, who already have enormous concerns about commodity price cuts and the resulting implications for their future viability. Also, the level of standardization and surveillance is very new to a group who have been used to escaping environmental restrictions and exerting enormous influence on the Irish state. There is a distinct air of indignance about their protests that matches their heretofore high status in Irish society. In this politically charged atmosphere, the Directive is portrayed as yet another nail in the coffin of the Irish farmer, as if this is an undifferentiated category.

Farmers are worried about the cost implications of long periods of slurry storage, the perennial issue of the bureaucratization of farming and the limitations placed upon their activities. Some farmers were interviewed in the *Farmers Journal* on the issue. These are some of their views on the current situation:

> It's the most intensive farmers out there who'll have most problems. We won't be able to expand in the future, we won't be in business. Yet we're being told all the time we have to expand.

> I'd like to progress in farming. I'd love to milk twice as many cows as I have now. But will the Nitrates Directive leave me any opportunity to do so? I can't expand now due to lack of quota. Not in future years ... the Directive could then prevent me from doing so, just as effectively.

These farmers have been led to believe for over a generation that a 'good farmer' expanded his/her production and made maximum use of the natural resources at his/her disposal. This was often at the expense of the surrounding environment, as we now recognize. From now on, however, being a good environmental manager is going to become part of the definition of a 'good farmer', and this may indeed produce some casualties among the farming population.

The Nitrates Directive has become an extremely divisive issue. Opinions differ on the matter and the usual factions emerged. The advocates of productivism are up in arms, while the environmentalists can now sit back and say 'we told you so'. The claims made on the countryside by green lobbyists now have official legal backing from the EU. It has united the IFA, Teagasc and Glanbia, who collectively view the limitations on farming activity as a major problem. All of these organizations are now conducting urgent talks as to how to move forward on the issue.[55] What is different this time, however, is that it appears that no amount of political posturing will alter the requirement that pollution will no longer be tolerated from any farmer who is in receipt of EU monies. The emblem of water pollution is the main driver of EU agri-environmental policy at this time. It has already been deferred for a long time and the EU Commission is now saying to the Irish state 'enough is enough'. A key point is that the law has been in place since 1991 and the only current change is that farm payments are to be tied directly to it from 2005, hence it cannot be avoided, as it was in the recent past. The 'bottom line' has been outlined by Martin Cullen, Minister for the Environment:

> Ireland has to comply with the court judgement handed down from the Commission and will do so at all costs. There is no way we can submit an action plan with greater than 170 kg/ha ... In fact, the debate that should have started in the mid-1990s has not emerged up until now.

The ruling by the European Court of Justice was welcomed by Anthony Waldron, an angler and member of Carra-Mask Water Protection Group in Co. Mayo. This group were instrumental in the case being taken in the first place.[56] Patricia McKenna, Green Party MEP, accused the IFA of trying to block governmental action on this piece of legislation. The IFA President, John Dillon, stated that the 'Nitrates Directive had been overtaken by significant improvements in water quality', and that 'the only winners are those who have nothing to do with farming'.[57] The IFA in Waterford, a key dairying county, plan to boycott the services of Teagasc, because they are recommending farmers to cut down on their nitrogen use, i.e. simply to prepare for the inevitable. In response, Teagasc have emphasized that they are not responsible for drawing up policy, but just have a consultative role in the process.[58] Outside the main factions, the more independent and informed voice of Matt Dempsey, editor of the *Farmer's Journal,* writes:

> If farmers are to be deprived of some of the money to which they will be legally entitled from next January, then it must be on the basis of clear, legally-identifiable infringements. The Nitrates Directive needs to be founded on sound science and sound implementation. Both are lacking in the current model.[59]

He has clearly recognized the inevitability of current policy trends, and instead of fighting it, emphasizes that the rulings should be implemented in a fair and transparent fashion. These changes are difficult for larger farmers to accept, as were quotas in the eighties. Some may try to fight the Goliath of the European Commission, but rather, it is inevitably those who focus their energies on adapting to the new policy climate and changing consumer demands who face the most promising future in farming.

Conclusion

The greening of Irish farming has produced several interesting results. Up to now, it has led to the zoning of the Irish countryside. REPS has been applied by those farmers and regions previously termed 'non-development' and who were thought to be somewhat backward. They are now being consigned to the function of providing environmental goods and primarily thought of as custodians of the land rather than as producers of food. They can no longer simply use the land as they choose, to maximize their own profits, and are becoming publicly accountable in new ways and caught up in a nexus of state control. Peasant farming is now also being packaged in a consumer-friendly fashion, primarily to sustain the burgeoning tourist industry. REPS is a mere nod in the direction of sustainability, however, in an era when the incentives for agricultural intensification are still very much intact.

This approach may be described as green capitalism because intensive farmers in wealthier regions have been allowed to continue with their productivist practices, which are much more likely to pollute the environment. Those farmers who are relatively well-endowed with economic, cultural and social/symbolic capital have so far had the ability to prevent such environmental restrictions being placed upon them. Neither REPS nor any other environmental sanctions have so far applied to those productivist farmers who are charged by the EU and the state with the bulk of food production in Ireland. This has now changed, however, with the tying of the SFP to cross-compliance regulations. Larger Irish farmers have not taken this lying down. They have continued to resist by querying the scientific basis of the Nitrates Directive, the rule that effects them most. There is currently a climate of fear among farmers and they see this as yet another way for the EU to make their lives difficult, if not impossible. The next chapter examines some other conflicts that are currently occurring in the Irish countryside.

chapter 5

Landscape and Heritage: The Politics of Perception

The perception of a tract of land depends completely upon one's relationship with that land. When looking at a bog, for example, a farmer may long to drain it so that it could become nice and tidy, ready to grow grass, a biologist will view it as a treasure-trove of biodiversity or an archaeologist might locate some ancient organic remains that enlighten us about the material culture of ancient times. The farmer has to make his/her living off the land, while the latter two are interested in keeping it ecologically pristine, for aesthetic and scientific reasons.

Like broader landscape issues, the preservation of Irish cultural heritage is also a hotly contested terrain. 'Heritage' is now very broadly defined by specialists in the field to include the physical landscape, with the ecological habitats and the flora and fauna that inhabit them; and our cultural practices like the Irish language, food, music and sport. This is as well as the more conventional understanding of heritage, that of archaeology, historical buildings and vernacular architecture, i.e. the cultural imprint of humankind on the landscape over the millenia.

This chapter is an investigation of contemporary attempts by both the Irish state and the EU to conserve valuable parts of Irish landscape and heritage, and the ideological battles that have ensued as a result.

Landscape: Ways of Seeing

This book has so far has been all about people, whether EU bureaucrats, Dublin civil servants or Connemara hill farmers. When dealing with rural issues, a key ingredient is of course the land itself, which comprises the places that are affected by human action and interaction. A place is made up of people, social actors, interacting with the physical environment. This of course also means that the physical reality of the place is changed by the social relations between these actors. The place is not just the background to the sociological subject, but may be viewed actually as one of the actors who loses or gains in symbolic knowledge struggles. The land and physical environment itself is not an homogenous plane upon which actors compete for dominance, but it is a varied entity which exerts a power of its own over the uses to which it can be put.

One sociologist explained that he reads 'landscape as a formation of socially constructed and historically determined visual signs and discourses'.[1] To translate this into everyday language, he means that the way the landscape is managed depends upon the actions of the people that have been responsible for it. These people look at the landscape in different ways that have been passed on to them from the society and culture in which they live. Certain ways of visualizing and talking about the landscape become normal in society, and others marginalized. These become inscribed in the culture, influencing future generations. Underlying this process lie power relations, because some people have the wherewithal to have their ideas dominate over others. All of this varies dramatically over time and across space. What is acceptable and 'beautiful' at one juncture is seen as objectionable in another. The landscape is therefore an archive that reflects the collective memory of people and nature, past and present. Uncovering its secrets allows us to interpret history and to decide upon the best means of interacting with the land for the benefit of future generations. This analytical process naturally produces different versions of the truth regarding the landscape. Competing constructions of environmental problems are 'constrained by and channelled through existing structures of economic and political power'.[2] This implies that competing sets of ideas about the environment need to be analysed within the context of existing sociopolitical structures. According to a constructionist perspective:

> What is ultimately most significant here is the process through which environmental claims-makers influence those who hold the reins of power to recognise definitions of environmental problems, to implement them and to accept responsibility for their solution.[3]

Another way to say this is that different sets of claims about 'what the problem "really" is"[4] compete for dominance. This approach leads us to an analysis of power relations in this arena, trying to to find out why one version of events is believed in official circles and another is not. A clear example here is the way that environmentalists' views were adopted by the EC in the eighties because their goals synchronized well with the goal of cutting the funding that went to agriculture. Economic efficiency was thus framed in ecological terms.

A social constructionist approach to rural and environmental social research also implies an anti-objectivist stance. It is assumed in conventional science (of both the physical and social varieties) that the researcher can remain objective and aloof from the subject of his/her analysis. In contemporary social science, it is common to reject this objectivist approach outright, and to acknowledge that personal bias can enter research at any stage. Using this more modest perspective, scientific knowledge is recognized as a social construction laden with a whole set of cultural and political baggage. As Bird advises:

> scientific knowledge should not be regarded as a representation of nature, but rather as a socially constructed interpretation.[5]

While elements of the environment are measurable, the results of the measurements conducted depend on who does the measuring. Following on from this, the public's understanding of the physical world depends upon what they are told by those who do this measuring. Social factors are extremely important in determining who says what ultimately gets done with the physical environment.[6] Objective science is not seen as inherently more valuable as a starting point or necessarily closer to the truth than that of a non-expert.[7]

The leading scholars of both the physical and social sciences are no less eligible for scrutiny in this regard. For example, Frank Mitchell (who wrote *Reading the Irish Landscape*) was a trailblazer who helped people to understand the palaeontology and physical structure of Ireland. He was a pioneer, mainly because he was an Irish scientist for an Irish public. His was a biocentric view, however, which prioritized nature over humans. From this perspective, the human impact on the landscape is most often seen as a problem rather than an asset. Others focused more on the human impact on the land, which is of more interest here. For example, one highly influential study of Irish rural life was *Irish Folk Ways* by E. Estyn Evans. Evans mined the Irish countryside for the ordinary ways that people worked the land, providing an unprecedented amount of detail on the material culture of small farmers. He argued:

> The history of rural Ireland could be read out of doors, had we the skill, from the scrawlings made by men in the field boundaries of successive periods. In them the unlettered countryman wrote his runes on the land.[8]

While this was no doubt a valuable exercise at the time, a retrospective look at his scholarly approach exposes it in a less positive light. Evans chose to treat Irish rural life as a research laboratory, focusing on the quaint farm implements and practices that could no longer be found in 'most parts of Great Britain'. With such a comparison on the first page of the book, it is no surprise that his perspective was a rather condescending one which could be read as an attempt to reinforce a romantic image of Ireland as a haven from modernization. It was less a critique of the dominant modernization perspective at the time than an escape from it, leading his followers down a political cul-de-sac. The apparently benign approach of microscopic examination of the tools used by Irish peasants distracted a whole subsequent generation of students and scholars of Irish rural life from asking more penetrating questions,[9] such as why things were the way they were, what the underlying social structures were that maintained Ireland's difference from 'Great Britain'. Evans was writing in the fifties and sixties, a time when people in the Irish countryside were struggling to cope with cataclysmic social change, resulting from increasing urbanization and social differentiation, crippling poverty and massively high levels of emigration. The resounding silence on these issues from such a leading scholar is baffling to say the least. The following observation certainly seems to describe the romantic, yet detached gaze that clouds Evans' work:

> The bourgeois perception and representation of the countryside as picturesque idyll or sublime spectacle has served to obscure the private property relations and forms of social domination which in capitalist societies have actually shaped the lie of the land.[10]

The social detachment of scholars like Evans left a vacuum to be filled. Anybody interested in the social reality of the time in rural Ireland had to go instead to poets, novelists and playwrights for enlightenment. Writers such as Patrick Kavanagh, Edna O'Brien, Kate O'Brien, Patrick McGinley, John McGahern, Brian Friel and John B. Keane ably coloured in the detail of the reality of country life for eager audiences starved of social critique.

Since Evans' time, rural Ireland has changed almost beyond recognition. The natural world has been tamed, cultivated and polluted at the behest of the state. The productivist 'story-line' enjoyed supremacy for several decades and it is really only now that a serious challenge is being mounted against it, with the advent of decoupling, the SFP and cross-compliance. In just a few years, the idea that farms now constitute 'landscape' has become predominant. Rural landscape is like a 'flickering text … whose meaning can be created, extended, altered, elaborated and finally obliterated by a touch of a button'.[11] The meaning that is attached to the natural world of soil, fields, trees, hedgerows, birds and wildlife has now changed in official EU circles. Ecological considerations have been promoted from the sidelines to centre-stage. Farm advisors and farmers are now on a sharp learning curve, just as they were in the sixties when they were just beginning to sow the seeds of the environmental problems that must now be solved. Back then, they had to learn how to intensify their production methods, and now they may have to re-learn some of the old ways of extensive farming.

The Designation of Environmentally Protected Areas

In Ireland, rural people have a strong emotional attachment to the land and an even stronger political attachment to the principle of private ownership. In this post-colonial society, the collective memory of fighting for land ownership is a potent force. As a result, it can be difficult for a farmer to accept the authority of an external force like the EU, which has transformed the 'land' into 'landscape'. Farmers' private control over their land now has to take a back seat to wider considerations of landscape conservation and the 'public good'. Polluting practices will no longer be tolerated and areas that harbour precious flora and fauna are set aside and designated for special protection.

The designation of certain areas as environmentally sensitive has been enforced by the EU since the mid-nineties and administered in Ireland by Dúchas, attached to the Department of the Environment (DoE). The official name for these is **Natural Heritage Areas (NHAs)**, which contain rare wildlife habitats and endangered species of flora and fauna. By the late nineties, 1200 of these had been proposed, covering about 750,000 ha.[12] It is only since December 2002 that the first batch of 75 of these have been granted statutory protection. These are all raised bogs (mainly in the Midlands), which are among the most distinctive types of Irish landscape feature. This designation is supposed to be incorporated into each county's Development Plan, and no grants are to be provided for any damaging developments.[13]

A second type of conserved area are **Special Areas of Conservation (SACs)**, an EU scheme purporting to satisfy the legal requirements of what is known as the Habitats Directive. More correctly, this is Directive 92/43/EEC (of 21 May 1992) on the conservation of natural habitats and of wild flora and fauna. This deemed that it was 'necessary to designate special areas of conservation [SACs] in order to create a coherent European ecological network according to a specified timetable'.[14] As of the end of 2003 there were 486 of these throughout the country. These are prime wildlife conservation areas chosen from among the proposed NHAs. The Habitats Directive lists certain habitats and species of flora and fauna that must be protected with SACs. These include the following:

Habitats	Species
Raised Bogs	Otter
Machair	Lesser Horseshoe Bat
Turloughs	Bottle-Nosed Dolphin
Blanket Bogs	River Lamprey
Bog Woodlands	Freshwater Pearl Mussel
Coastal Lagoons	Killarney Fern
Fixed Coastal Dunes	Marsh Saxifrage

Thirdly, another designation is **Special Protection Areas (SPAs)**, which arose specifically from what is known as the Birds Directive. This is Directive 79/409/EEC (of 2 April 1979) on the conservation of wild birds. This required the designation of SPAs and wetlands in order to protect biotopes and flora and fauna within two years of adoption. There are currently 110 SPAs countrywide, covering over 220,000 has of land, and the landowners on another twenty-five sites have been notified of designation in the near future. These are wetlands that attract migratory species and other places where rare or threatened bird species feed or roost. It is generally acknowledged that the health of the bird population is an indicator of overall environmental health. The title of the landmark publication *Silent Spring* by Rachel Carson (1963) refers to the deathly silence of a spring without birdsong, as they had been killed off by the over-zealous use of DDT. The decline in bird numbers therefore served as a pointer toward monumental environmental problems. While the use of DDT has since been outlawed, there are threats from plenty of other exploitative farming practices that destroy or threaten bird habitats.

Bird diversity adds an incalculable quality to the landscape. The beauty of even the most common birds, when studied in detail, makes it incumbent upon us to do our best to preserve their habitats and encourage their proliferation. Even among those who do have some affinity for birds, it is common to refer only to the rarer species, to the neglect of the more common ones. The more colourful and dramatic birds are, the more likely we are to notice them.

To complement the work of the state officials, BirdWatch Ireland is the main Irish non-governmental organization who lobby for the conservation of wild birds and their habitats. They conduct extensive surveys of birds that feed back into conservation policy in Ireland. They also manage nature reserves, where birds can feed and breed without threat. They are funded by membership fees and thus manage to organize lectures, courses on wildlife and guided field trips.[15] Together, BirdWatch Ireland and the Royal Society for the Protection of Birds (RSPB) in Northern Ireland have agreed on three designated 'lists' of bird species.

The **Red List** are those of high global conservation concern, whose breeding population or range has declined by more than 50 per cent in the last twenty-five years. Many of these are already virtually extinct. Among this list of eighteen species are: Barn Owl, Nightjar, Chough, Corncrake, Yellowhammer, Curlew, Hen Harrier, Corn Bunting and Red Grouse.

The **Amber List** has seventy-seven species, and these are defined as rare or sporadically breeding species, whose breeding population has declined by 25–50 per cent in the last twenty-five years. What used to be some of the most common birds are here: Swallow, Kingfisher, Sand Martin, Cuckoo, Puffin, Peregrine, Skylark, Snipe and Cormorant.

The **Green List** is what is left over after the preceding ninety-five species are eliminated, and whose conservation status is considered favourable. The loss of such diversity among birds is one of the hidden costs of commercial farming, where ditches are removed, pesticides and fertilizers used liberally, and old farm buildings demolished. Modern farming methods therefore destroy breeding habitats, limit nesting opportunities, remove cover, and reduce food availability, making it difficult for these fragile creatures to survive.

In addition to the above designated areas, seventy-six Statutory Nature Reserves have been listed in the Republic. These are meant to protect rare species of birds and animals. There are also six National Parks, which cover an area of 47,500 ha. These are: Killarney National Park, Co. Kerry; Glenveagh National Park, Co. Donegal; Connemara National Park, Co. Galway; Wicklow National Park, Co. Wicklow; The Burren National Park, Co. Clare; and Owenduff-Nephin Beg National Park, Co. Mayo.

Finally, there are fourteen Biogenetic Reserves and two World Biosphere Reserves, which emerge from the Council of Europe and UNESCO, respectively. Together, all of these designations comprise what is known as NATURA 2000, which is a coherent European network of the best nature conservation areas.

So how does this process work? How does a region get to be designated as a protected area? The official booklet provided by the DoE, *Living With Nature*, describes the process as follows. NHAs, SACs and SPAs are chosen based on existing published research by professional naturalists and also using inputs from amateur ecologists and relevant NGOs. DoE scientists then go to evaluate the area. Upon designation, landowners are then notified that the area has been proposed and they are told what measures are required to protect the site. Farmers are usually required to sign up to REPS and no farm practices that damage wildlife will be tolerated. As well as this, the proposed areas are advertised in local newspapers and on radio. Landowners can make objections or appeals to the DoE, but these must be made on scientific grounds only. If the submission is complete, the area will then be re-examined and the matter discussed with the landowner. If this appeal fails, the landowner can take it to the next stage, which is the Nature Conservation Designation Appeals Board, an independent body made up of both landowners and conservationists. The DoE says that regular consultation occurs at national level with farming and conservation groups. At local level, landowners are encouraged to form Liaison Committees to negotiate with the DoE.

If one were to end one's research here, taking the government at their word and asking no more questions, one would think the future of Ireland's biodiversity was in safe hands. However, the excruciating slowness of the implementation of the crucial EU Directives has caused some trouble for the Irish government. In July 2004 the EU Commission initiated legal proceedings against Ireland in the European Court of Justice for failing to implement EU environmental protection law. There were seven cases of legal breaches regarding habitats, sewage treatment, illegal waste dumping, water pollution, wild birds and waste management, and two cases of failing to comply with previous rulings.[16] Six years ago, back in 1998, the EU Commission highlighted that Ireland has one of the worst records in Europe on nature conservation,[17] and many of the same sites are still a major problem, such as the landfill site at Tramore, Co. Waterford.[18] When the breaches were made known to the press, the then Minister for the Environment, Martin Cullen, responded:

> Ireland is 98.8 per cent compliant with EU law ... Yes, there are matters outstanding and the government is proactively working with the Commission to deal with these. Just as in all member states, challenges will occasionally arise.[19]

The main challenge they have to deal with clearly is the EU breathing down their necks. One area that was singled out for attention was the Owenduff-Nephin Beg region of Co. Mayo, which is the country's largest SPA. The plans to reduce sheep numbers to prevent overgrazing have not been fully implemented, hence habitat damage to previously heather-covered hillsides is ongoing. There has been a general failure to put in place measures to properly protect rare species such as the Lesser Horseshoe Bat and the Natterjack Toad. If this non-compliance continues, it is said that the Irish state could be facing daily fines of €20,000.[20] The everyday dealings that Dúchas staff have with farmers have led to some minor revisions of SAC boundaries. Conservationists are not happy about this and the EU is angry because of how slow this makes the whole process of implementation of environmental protection.[21] The snail's pace of the EU bureaucratic machine has often been criticized, but it is outpaced only by the even slower implementation of its most valuable legislation in its westernmost member state.

The criticisms from the EU will have to be taken seriously by the state eventually, but what about criticisms at home? The main conservation organizations (An Taisce, Irish Wildlife Trust, BirdWatch Ireland, Irish Peatland Conservation Council and Coastwatch Europe) consider Dúchas's list of designated sites to be too limited and they submitted an extra list of sites in 2000.[22] Conservationists argue that the SACs and SPAs would be 'left as ecological "islands" in an inhospitable, humanised landscape of intensive farming, roads and rural suburbia'.[23] When certain areas are highlighted as being ecologically important, it may lead to viewing what is 'in-between' as less important.[24]Along similar lines, Aalen argues:

> The proliferation of designations encourages a compartmentalised mentality which disaggregates the landscape, treating it as a series of disparate parts rather than as an intricate unified system of interacting elements.[25]

While Aalen argues for a holistic approach to conservation, Feehan argues that everybody needs to develop an appreciation of the ordinary beauty of what is in their own patch, rather than necessarily heading straight for an SAC to appreciate 'nature'. He argues passionately:

> The nurturing of biodiversity at local level may not seem particularly important from a national perspective. But it is extremely important at community level, because *this is our bit of the world*. It is through Nature as experienced on this level that our wonder and understanding and appreciation are nurtured … The preservation of rural biodiversity on a local scale affects all the community; its loss or decline diminishes the richness of all our lives and lessens in value the inheritance with which we endow the next generation[26]

Also, from a pragmatic perspective, ecological stepping-stones and corridors need to be protected. This is so that wildlife can have the necessary amenable routes through which they can travel and expand their gene pool.

This expert discourse expressed by conservationists does not necessarily make them too popular. John Feehan, a leading conservationist himself, is only too aware of this. He says that in the rural mind nature conservation is often associated with a privileged élite with plenty of leisure time, with which they can spend doing trivial things like recording wildlife. This is a hangover from the nineteenth century and is often still applied to groups like An Taisce. He writes:

> The identification has survived to our own time, and has been one of the most stubborn of all obstacles in the campaign to educate the community to an environmental consciousness.[27]

As an example of this, one English woman who was interviewed for this research is active in An Taisce. She knows that the organization is unsavoury to the long-term rural residents:

> The trouble is it's difficult to get local people to put themselves in important positions and I can really understand that. There's a lot of family pressure and from people they know – it's not easy for anybody to stand out from the crowd. Therefore we're an easy scapegoat if you like, English people or foreigners, you have got a lot to lose but in some ways, there's a lot of things to lose that you're never going to gain in the first place. It's easier for us.

Tovey argues that the NHA conservation strategy emphasizes the 'scientific gaze' on the countryside, that it 'takes for granted that the environment as it appears to geologists, botanists and zoologists has a reality which is more objective, or more valuable, than the environment as constituted in farming or in the daily lives of rural people'.[28] She argues that the conservation strategies used in both the EU and the Irish state appear to view farmers as inconvenient inhabitants and indeed exploiters of the land. They are viewed as 'surplus populations'[29] who should consider finding employment in another sector and leave the land either to be farmed by commercial efficient farmers or landscaped by botanists and conservationists. Feehan argues that the designation of NHAs is similar to the mapping of the land by the Ordnance Survey in the nineteenth century, in that it officializes and centralizes the meanings of places, taking away control from local people.[30]

And what of the critiques from these local people? What are

the experiences of farmers, landowners and other rural dwellers on this matter of designation? Over the years, there seems to have been widespread confusion regarding the whole business. Under REPS I, if a farmer had any small portion of land in an SAC, s/he got a top-up payment (of c.€15) for the entire farm (max 40 has). Since REPS II began in 1999, the top-up payment has only been paid for the portion of the land that is actually in the SAC. There was a lot of resentment about that at the time, as seemingly arbitrary rules were being imposed on a rural population who badly needed the extra income. The IFA Western Development Chairman, Gearóid O'Connor, warned in 1998:

> Failure to win approval for the entire package puts in doubt acceptance by farmers of the environmental designation and restrictions imposed on their land.[31]

The potential restrictions on their farming activities is a major source of worry to farmers. Some commercial farmers in north Co. Limerick had severe misgivings when they were told that their land was to become part of an SAC. Even though the official line from the DoE was that consultation occurred where possible, these farmers' experiences were very different. They were very worried about the future:

> We have heard that it's possible there may be no manuring, no spraying, and no silage cutting in the future or that silage won't be cut until late July. If this happens, farmers will be in serious trouble. We would need substantial compensation for these changes.

> I sell reed for thatching that I cut myself. In fact, I invested in a £10,000 cutting machine just two years ago. The reed is a good income earner for me. What will I do if that is no longer allowed?[32]

There seems to be a knowledge vacuum here, where farmers have difficulty getting answers to these basic questions. One local organized opposition to SAC designation occurred in Lady's Island, Co. Wexford, where local people set up an SAC Awareness Committee. One member of this group is quoted:

> Yes, we need to protect wildlife but are we going to protect it to such an extent that the livelihoods of inhabitants living close to sites of scientific importance are negatively affected? Which comes

first, man or beast? ... It comes down to your rights being totally trodden upon. Right now there's the feeling that wildlife is first class and we're second.[33]

Because the designations are so 'top-down' in character, farmers can often see little point in changing their practices, on which they depend for their living, to protect perhaps a rare variety of snail or moss. People can easily see the point in preserving creatures higher up the ecological ladder, like the bottle-nosed dolphin, but perhaps do not empathise with a river lamprey, which isn't quite as pretty! Michael Viney, the naturalist, argues that 'the whole conservation apparatus ... has been presented to a rural society generally so ill-equipped to see their point'.[34] He goes on to say that it is lucky that many species happen to live in areas that are scenically attractive and most people can see the social and economic value of conserving these. Promoting an abstract value like the need for maintaining biodiversity among a group of people who make their (unreliable and intermittent) living off the land is a fraught exercise that is bound to cause conflict from time to time, especially in the buffer zones between city and country.

Rural Tourism and the War about Walking

The topic of landscape conservation cannot be separated from that of tourism, which is one of the most significant indicators of the globalization process. An enormous and ever-increasing number of people travel all over the world every year for their holidays, in search of unique and exotic experiences. The surge in consumerism in recent decades has led to the packaging of places like rural Ireland in ways that purport to please the 'tourist gaze'.[35] The ongoing search for the tourist dollar tends to create social groups who have a direct monetary interest in bringing visitors to different parts of the country. Sometimes this form of socio-economic activity is carried out in a way that adds diversity and enhances local social life, but sometimes it also causes conflict, pollution and congestion. It depends on whether the preferred type of tourism facilitates the appreciation of the environmental and cultural wealth of rural Ireland. At one end of the spectrum of possibilities lie independent travellers. They are more likely to spend time in a place, talk to local people, and 'give more back' to an area, both emotionally and economically, than the mass tourist. At the other end of the spectrum is the latter type, who will do a whistle-stop group bus tour of the country in a week, 'doing' Dublin, Killarney and Galway, enabling them to tick Ireland off their list of countries. There are of course other possibilities in between these extremes.

Certain areas of Ireland, especially in the west of the country, have been tourist destinations for a few centuries. The wild and rugged hillsides of Connemara are the epitome of a 'picture-postcard' rural image that has had a resonance for millions of international visitors. This popularity has focused the minds of planners. It is no accident, for example, that the priority areas for concern by EU environmental scientists have been the key tourists sites of the National Parks. Ironically, tourism can actually contribute to the generation of more concern for the quality of the environment. It was in one of the National Parks, the Burren, that we witnessed an impassioned campaign regarding the construction of tourist facilities and an interpretive centre.[36] Certain sites attract huge numbers. The following are some visitor numbers in prime areas for 1996:

Connemara National Park: 70,000
Glenveagh National Park: 82,000
Aran Islands: 100,000
Killarney National Park: 200,000[37]

Some will be happy to see these numbers, while others might dread their environmental impact. The wild beauty of parts of rural Ireland can no longer be taken for granted, as there are several different types of social forces that can contribute to their deterioration, from the suburbanization of rural towns, to overgrazing by sheep to mass tourism in areas like the Ring of Kerry and the Aran Islands. The rural environment has become one of the key stakes in the redefinition of contemporary Ireland. Nature is consumed by tourists in different ways, as they pick and choose environmental elements. Catering to tourists in rural areas is considered by the state to be an important means of creating jobs and tackling rural poverty.[38] Whether or not this works out in reality is another question, to which there are few answers.[39] There is not the space here to go into detail on rural tourism. Instead, a focus will be placed upon one issue that has animated the rural public, especially in the west of Ireland: that of access to the countryside by hillwalkers.

This issue was brought into the public eye in early January 2004 by the jailing of a Co. Sligo farmer. Andy McSharry was accused of intimidatory behaviour by a couple who had been walking on his land. He had taken photos of the couple, which he claimed were to prove that they hadn't been injured while on his patch. He refused to pay a court-ordained fine, so he was sent to jail. The IFA supported him in his actions. Living on the picturesque slopes of Ben Bulben, McSharry has had to tolerate walkers for many years, especially since local tourist enterprises mistakenly advertise part of his land as a public walkway. McSharry also charges walkers for access to his land.[40] A letter writer to the *Farmer's Journal* is very upset by McSharry's actions. Michael Gibbons, of Connemara & Islands Heritage Tourism Ltd. argues:

> Notwithstanding the necessary and honest debate over access to the land, men like the 'Bull' McSharry have nothing to contribute except to lead Irish farming down a cultural and economic cul de sac ... The road to conflict is the road to ruin.[41]

Because this letter-writer is directly involved in the tourist industry, he does not like to see such conflicts emerging.

The crux of this conflict involves public liability insurance. Currently, if one falls and hurts oneself while walking across private land, one can sue the landowner for damages. The way around this is that there can be marked walkways that serve as rights of way and hence are exempt from this law, and walking clubs can carry their own insurance to cover their members. There are currently only 3000 kms of these marked ways in the Republic, compared to 225,000 kms in Britain. In Britain the law has enshrined the 'right to roam', which means that walkers have access everywhere on the land, including farmyards.[42]

This is very different in the Republic of Ireland, where landowners have the right to evict walkers from their land.

One such event occurred on 19 September 2004 in north Co. Wicklow. Three hundred people, including several TDs, went on a protest walk on what they claimed was an ancient right of way. The protesters used a 1757 map of the area to show that it was an old coach road, but this was disputed by a local landowner who claimed that it was private property and hence off-limits. A confrontation ensued involving the gardaí and the story made the national news.[43] In areas like this, where the urban and the rural are cheek by jowl, heated encounters like this are becoming more common, as the issue of rural access becomes more politicized.

The organization Keep Ireland Open (KIO)[44] has been lobbying for several years for improved access to the countryside. A key concern of theirs is the erection of sheep fences and the division of commonage. They want to see these practices banned and the land opened up to more walkers. They are angry about the number of signs they encounter on farmers' gates that ban access to their land. They argue that Ireland is losing tourist business because of this, at a time when farm diversification should be the order of the day.[45] In a letter to the *Irish Times* on 16 April 2005, their secretary pointed out what he saw as the difference between Ireland and unnamed 'other European countries', where 'landowners have some sense of community and, more importantly, realise that they owe something to the taxpayers without whose generous support they would not be able to live in a pleasant rural setting'. The tone of this message is: it's payback time! The vociferousness of groups like KIO probably signify to farmers how they are losing control over their rights over their own private land and are being railroaded and intimidated by urban-based interest groups.

Is there any answer to this open conflict between two groups who have directly conflicting interests? In response to the above debacle, Eamonn Ó Cuív, Minister for Community Rural and Gaeltacht Affairs, announced that he was to establish a council to deal with the issue, and to find some sort of common ground – Comhairle Na Tuaithe (Countryside Council). A promising attempt is being made at local level in Co. Wicklow to address the problem. Because of proximity to Dublin, the biggest population centre, lots of walkers make their way to the Wicklow mountains, especially on weekends. EU rural development money is to be used to negotiate between farmers and walkers in the county. The Wicklow Uplands Council, a voluntary alliance of thirty-five organizations, welcomed the publication of a code for countryside use by Agri-Aware. The latter is an organization set up to promote a positive image of farming and the food industry.[46] The key here seems to be real consultation between the parties, a clear set of rules regarding good behaviour and a little bit of goodwill on both sides. One local farmer states his position:

> We are happy to continue to let people walk so long as we have control. Both [my wife] and I are walkers and we find that most walkers are fine, decent people, but there's an element that is only interested in causing grief.[47]

The solution proposed by the Uplands Council is that landowners could lease access routes for a fixed period. The council could then maintain the routes and it would also cover the costs of insurance. It is too soon to say how this will work out, but it is certain that consultation is the only way forward on this issue. Local people need to feel they are involved in the negotiation process and the concerns of walkers' organizations also need to be considered. This is one issue that has the potential to deepen the already-established urban-rural divide in Ireland, so it is hoped that an amenable solution package can be found.

Haggling over Heritage

Ireland is full of important historical sites, from the 5000-year-old Stone Age remains visible at the Céide Fields in north Mayo, to abbeys, castles, cathedrals and churches that date from more recent centuries. As well as these are crannógs, dolmens, fulacht fiadh, passage graves, ringforts, round towers, souterrains, standing stones and tower houses. These remains and buildings tell the stories of the many and varied influences that have shaped twenty-first-century Ireland as we know it. The conquests by the Vikings, the Normans and the British are all etched on our landscape in a myriad of sites throughout the country, as is the powerful and enduring influence of the Catholic Church. Taking time out to visit these places injects a little magic into one's life, when we can mull on life in times past or just get some peace in an otherwise busy day. A beautiful castle or abbey can be like an oasis of calm in the suburbanized cultural and aesthetic desert that much of the surface area of this small island has become.

Archaeologists recognize the contradictory forces at work in contemporary Ireland with regard to heritage. Unprecedented wealth and concomitant growth, or what is often unfortunately referred to as 'development', has put the landscape under strain like never before. A record number of houses are being built (see chapter six) and a record number of new cars have been bought, leading to the privatization and urbanization of large tracts of what was previously farmland. As we have already seen, the intensification of farming is also an ongoing trend. At the same time, the policy on heritage management has taken a holistic turn. It is now recognized that it is not enough to simply view the archaeological sites in isolation, but that a total, holistic landscape approach is required.

This incorporates a wider concern about the landscape surrounding a round tower or a crannóg, rather than just the construction itself. There is therefore a consideration of the overall integrity of the countryside itself as an historical resource, including all of its aspects. So an 'archaeological landscape' is more than just a concentration of sites, as

> it is the interconnections between the components, whether those connections are chronological, spatial, social or functional, that provide additional information … the space between the monuments … are fundamental to an understanding of their import and integrity.[48]

This is obviously a much more challenging approach to heritage management, as it concerns all of the influences, both positive and negative, on the landscape. Logically, the adoption of this approach means 'taking on' a wider variety of social actors and interests, from farmers to builders and property developers. The drainage of land, the knocking of field boundaries, road building, tree planting, turf cutting and pipe laying: all of these activities are potentially very dangerous to the archaeological landscape. Between 30 and 60 per cent of monuments have been removed since the nineteenth century, particularly over the last thirty years.[49] This is obviously lamentable when one considers how immeasurably unique these features are.

As in the previous discussion of landscape conservation, if one trusted the state to take care of our heritage and one simply behaved like a tourist, flitting through the countryside, we could stop this chapter here. However, it has become clear that much of our heritage is under grave threat from road development and rampant suburban growth. The unprecedented demands being put upon our landscape means that we can no longer take for granted that our heritage features will be cared for as they deserve. For example, Frank McDonald points out that the National Monuments Act was amended in 2004 to allow the Minister for the Environment full discretion in any case, up to and including the destruction of a monument.[50] That a government minister could be accorded such power over irreplaceable and priceless artefacts is nothing short of astounding. It is poles apart from European policy, which advocates that archaeologists should be core participants in the planning process. This is provided for in the European Convention on the Protection of the Archaeological Heritage, 1992. As we have already seen, the best intentions of EU policy makers are often overtaken by the 'business-as-usual' approach of the Irish state.

The main organization entrusted with taking care of a broadly defined Irish heritage is the Heritage Council, established as a statutory body under the Heritage Act in 1995. Its role is 'to propose policies and priorities for the identification, protection, preservation and enhancement of the national heritage'.[51] It has a particular function in collecting data on Ireland's heritage, promoting public pride therein and proposing policy to the state on heritage issues. Its expertise is called upon when there is a controversy, for example, in the recent debacle surrounding the routing of the M3 motorway near the Hill of Tara, the seat of the High Kings of Ireland and one of our most ancient and precious monuments.

The National Roads Authority has chosen a route through the Tara-Skryne Valley, north of Tara, and this has been approved by An Bord Pleanála. However, a campaign group called Tarawatch objects to this route, saying that it is too close to such a valuable monument, and they have proposed an alternative route south-west of Tara. The Minister claims that a lengthy consultation process took place in the local area, and that it had public support, especially considering that it was farther from the Hill of Tara than the existing N3 road.[52] On the other hand, Tarawatch commissioned a telephone survey of 1000 people that found that 70 per cent of people wanted the motorway moved from the proposed route.[53] It is clear that there is a battle for the hearts and minds of the Irish public on this issue. The Heritage Council's statement on the plan said:

> Given that the responsibility of the Council relates to the National heritage, Council recognises the sensitivity of any decision with regard to the road location. If Council were the body with decision making powers on this issue it is most unlikely that it would have chosen this new route ... given the international significance of Tara it is a matter of debate if sufficient weighting was placed upon heritage in the matrix of criteria used to inform the decision making process.[54]

Frank McDonald maintains that 'because of its 40m width and the much larger interchange planned for Blundelstown, just 1km to the north, it will be impossible to conceal it in views from the Hill'.[55] Fintan O'Toole decries the development, and it represents for him cultural impoverishment, a loss of values and the 'uglification of Ireland'. He goes on to say:

> The M3 is not just an example of stupid and unnecessary vulgarity, of what seems a perverse desire to do everything in the ugliest possible way ... If the M3 in particular goes ahead in its present form,

the advent of a public culture that is incapable of recognising any values beyond the immediate and pragmatic will be undeniable.[56]

Dick Roche points out that none of the academic objectors challenged the original proposal in court in 2003. He claims that he is adopting 'a measured approach' that simultaneously protects heritage and provides new roads. There is clearly major and fundamental disagreement between the state and varied objectors to this plan. State representatives claim that they have been democratic and reasonable, while objectors would reject this outright.

Vincent Salafia,[57] an environmental activist, was given leave by the High Court to take action against the Minister for the Environment, Dick Roche, regarding the protection of the archaeological heritage in the area. This week-long hearing took place in January 2006 and the High Court decision is to be announced on 1 March 2006. It will be interesting to see the outcome in the High Court and whether the case made by the objectors will be strong enough to stop, or even park, the speeding juggernaut that appears to be (both literally and metaphorically) ploughing through our everyday lives and our imaginations in contemporary Ireland.

Conclusion

People generally only get exercised about landscape and heritage when their own piece of territory is challenged by some outside force. They can live their lives for years without consciously thinking about what surrounds them, until a competing claim is made on their immediate environment. The Irish landscape and the heritage it represents has been subject to the dictates of productivism for decades. This of course continues, but larger tracts of land are now being treated as 'landscape' and conserved for ecological reasons. Catering to the recreational needs of the urban population and also the urbanization of rural people mean that the fields, hills, mountains and waterways are being visualized in a way that is contrary to the

viewpoint of the conventional farmer. Many of the new conservation measures seem to be imposed in a top-down fashion,
perhaps despite the best intentions of Dúchas. Landowners
need to feel a sense of ownership of these new measures before
they will adhere to them. This could take time and more efforts
need to be made to draw all the relevant social actors into the
debate, or SACs and SPAs will just become something else to
resist and subvert. Nature's present and future is intrinsically
bound up with human activity of all kinds. The management of
that activity is a highly political arena. Without some vestige of
local democracy with regard to this issue, everyone will suffer
– farmers, taxpayers, government and, above all, Ireland's
besieged wild birds, plants and animals. I will leave the last
word on this subject to John Feehan:

> The future lies with an ever-widening appreciation of all the ways
> the countryside enhances human well-being and a determination
> to sustain it. And with a farming community that can bring the
> best in human ingenuity, intelligence and sensitivity to bear upon
> that end, and is able to make a decent living by so doing.[58]

chapter 6

Contrasting Rural Visions: Two Key Issues

The Irish countryside is composed of an increasingly diverse set of social actors. The meaning of 'rurality' is up for grabs by many varied groups and individuals. Conflicts over how the land (which is now often called the 'environment') is used form a key part of public discourse in contemporary Irish society. Debates on certain key issues tend to reveal exactly who the actors are in this arena. The two debates addressed in this chapter are those of housing in rural areas and 'clean' food production.

With more wealth in evidence following the 'Celtic Tiger' boom period, more people want their own little piece of the Irish countryside. This new ownership may take several different forms. Firstly, because of the over-development of the Dublin region, the 'commuter belt' now extends as far as counties Cavan and Carlow, putting unprecedented pressure on towns and villages in this extended region and increasing the amount of traffic on the already dangerously congested roads. This in turn puts pressure on the state to build more and bigger roads, creating a vicious circle that has yet to be tackled. Secondly, the new houses are often second homes, which now form a core symbolic part of middle class identity. It is not uncommon now to own a home both in the city and in the countryside, and also in France, Spain, or the now burgeoning regions of central and Eastern Europe.

Finally, the next generation of rural dwellers are also building new houses, so the combination of all of these groups are adding up to lots of new housing developments in rural Ireland. The demand for new housing is matched on the supply side of the economic equation by the retreat from farming, meaning that more farmers are willing to sell small plots of land for sites. The increasing concern for environmental issues in Ireland, both at popular and official levels, has led to the emergence of both ideological and indeed physical battlefields.

Another battlefield is centred around the politics of food. The apocalyptic sight of mountains of burning cattle during the recent Foot and Mouth Disease (FMD) outbreak in Britain served as a shocking testimony to the utter madness of conventional farming. This vast infernal chamber of horrors was the result of a farming system that treats nature and animals as mere economic units to be manipulated in order to maximize profits. Farmers were inconsolable as their herds were rounded up for slaughter. The so-called 'modern' system that produces vast quantities of such goods as beef, milk and grains, exploits nature and pushes it well beyond its natural limits. This has led to the production of food that is not fit for consumption by human beings. FMD is just one of a litany of food scares that have bedevilled industrial farming in recent years, from BSE to avian flu to E.Coli, among others. Farmers who adopt organic methods avoid such problems. The second half of this chapter addresses the prospects for the continued development of organic farming in Ireland at this time.

Housing in Rural Areas

The apparent prosperity experienced by many in recent years, the willingness of banks to lend money and of ordinary people to borrow it, and the widespread desire to own at least one substantial stand-alone house has led to an impending planning crisis in rural Ireland. The desire for a stand-alone house has been attributed by the eminent historical geographer, E. Estyn Evans, to the symbolic association of hamlets or clachans with the 'squabbling poverty' of the Famine era. He says nearly everyone wishes 'to have a house where he cannot be over-looked'. A similar conclusion was reached by a less scholarly, but perhaps equally perceptive author, Pete McCarthy, author of the bestseller *McCarthy's Bar*. He says:

> The self-confident synthetic consumerism of the new houses is a response to this history of deprivation. 'Look,' the bungalows are saying, 'we're not peasants anymore. We buy things now rather than digging them up. We've been sitting on bare wooden benches for centuries, but we've got Dralon sofas now, just like you.'[1]

Opinions are bitterly divided on the issue of housing, and each group often distorts the public debate in order to suit their own ends. Environmentalists may portray farmers in stereo-typical terms as backward polluters of the land, and farmers may portray environmentalists as sanctimonious city types who want to trample all over their rights. This debate is often presented as a pitched battle between rural and urban interests, the former presumed to be interested in young rural dwellers being able to build and live in their own areas of origin, the lat-ter presumed to be only interested in preserving the rural land-scape to the exclusion of building new houses. The picture, however, is not that simple. Localities are composed of many varying interest groups and individuals who have an influence over what gets done.

It is first necessary to look to the EU in order to chart Irish government policy on rural housing. The dominance at EU level of the key idea of sustainable development has had far-reaching consequences for many different aspects of social life in rural Ireland. For example, the aforementioned 1997 document *Sus-tainable Development: A Strategy for Ireland* also incorporated guidelines for controlling rural housing. It specified that houses should only be built based on local need and that account should be taken of their impact upon roads, services and the environment. Fianna Fáil TD and Minister for the Environment, Heritage and Local Government, Martin Cullen, stated in March 2004 that these guidelines had been implemented in an overly strict manner, and that they should therefore be overturned and superseded. People who 'have roots or links to the rural community and are part of or contribute to that community'[2] should be allowed to build a home in their own areas. He also ruled that the designation of an area as an SAC or SPA (see chapter five for detailed discussion) should not necessarily present an obstacle to rural construction. However, any new constructions should be 'subject to good planning practice'. He cites a progressive new publication from Cork County Council, the *Cork Rural Design Guide*, as the best example to follow. The local authorities and An Bord Pleanála, who are responsible for assessing planning permission applications, began immediately to implement the changes. In the same article, Minister Cullen asserted that the concept of rural development 'requires an explicit acknowledgement of the role that people living in rural areas have to play in supporting a dynamic rural economy and social structure'.

The IFA has welcomed these changes, but thinks they have not gone far enough, and they are concerned that 'applications ... are being assessed by planning professionals whose training, background and expertise are of an urban character'.[3] The political opposition were happy to point out that the minister's proposal was forwarded only weeks before crucial local elections, in which Fianna Fáil subsequently lost 20 per cent of their seats. It was interpreted by many diverse political commentators as a means of wooing the rural voter at a time when Fianna Fáil popularity was slipping badly. This results from Ireland's persisting tradition of clientelist political culture, where votes have often been exchanged for favours granted. The various ongoing tribunals serve as a sickening testament to the pervasiveness of this corrupt culture. This 'nod and wink' politics is severely at odds with the strictly bureaucratic, regulationist and politically transparent regime favoured at EU level. This amounts to a severe cultural clash, and a quality rural environment is currently one of the main bargaining chips to be lost or gained in this game of political chance.

The other key document that serves as a backdrop to this governmental about-turn is the National Spatial Strategy (NSS),[4] which was published by the current administration. This document is designed to be the Irish 'ten commandments' for balanced regional development for the next twenty years. It was a response to the overdevelopment of the greater Dublin region and the real or potential depopulation of some rural areas. As a counterbalance to the development of the major urban centres, the NSS aims to foster growth engines in the regional 'gateway' towns of Letterkenny, Sligo, Dundalk and the trio of Athlone/Mullingar/Tullamore. These towns were chosen to complement the cities because of 'their location, critical mass of labour, developed transportation, communications infrastructure and access to markets'.[5] The idea is that infrastructures, amenities, businesses, and most crucially, people, should be focused in these towns in order to bring some balance to the highly primate Irish settlement pattern; and that this should lead to less congestion in the cities (especially Dublin), less commuting time, better access to crucial services and an overall better environment. If there is a 'critical mass' of population in these centres, it would provide the labour and skills needed to make businesses thrive. Along with this, the document advocates building up Ireland's urban structure, to combat the urban sprawl that is threatening the physical and social integrity of the countryside.

Frank McDonald, environmentalist and journalist, wrote in 2001:

> it's no consolation that a National Spatial Strategy is being cobbled together at the Custom House, because there isn't an ounce of political will to implement it.[6]

In 2004 he argued that the recent proposals from the Minister directly undercut the recommendations of the NSS. He asked, 'How are [the major towns] going to develop if life is drained out of them?'[7] He points out that none of them have a population of over 20,000, and with the way being cleared for more one-off rural housing, 'we will end up with a land where towns never end and the countryside never begins'. In a recent attempt to accelerate government decentralization, Cullen cajoled civil servants with the idea that they could buy mansions in the countryside with the inflated prices they would receive for their Dublin homes. This surely indicates that the complete opposite of progressive planning policy and lofty ideals like sustainable development is actually driving government policy. While the government may recognise the necessity of such ideals in their heads, their hearts certainly don't seem to be in it. Even if the will was genuinely there, to align rural housing policy with the trio of tourism, agricultural and environmental policy is a monumental task. This is the case even on paper, never mind in practice. As we have already seen with regard to REPS, one can never assume the clear linear implementation of any policy, as it is affected by each set of actors that it encounters on its convoluted journey from design to implementation. Each encounter can produce unexpected consequences that may change or undermine the implementation of the policy. The ideals of the NSS are bound to be undermined by the continued construction of so many one-off rural houses, which are currently going up at the astounding rate of about 25,000 per year.

This is an especially strategic issue in Gaeltacht areas. If unlimited one-off housing is allowed in these areas, it will lead to the dilution or diminishing of the Irish language in the few geographical pockets where it is spoken in significant numbers. The designation of the Gaeltachts is to undergo review in 2007, so even from a purely machiavellian perspective, it is unwise to threaten their special status. Grants from the Údarás Na Gaeltachta often fill a development gap in relatively impoverished areas. The employment created in Údarás-funded small businesses can be a vital lifeline for rural-dwellers, and can reduce dependency on multinationals, for example, who may not have the long-term endurance that a smaller company can. So the diminution of the lucrative culture of Gaeltacht areas can have long-term deleterious effects upon their economic well-being. Since they are in receipt of funding, which amounts to positive discrimination for Irish-speakers, it is surely in everyone's interest to maintain their integrity in this regard.

As already mentioned, water pollution has also been attributed to the exponential increase in rural housing and the well-established pattern of ribbon development. Each new house will have a separate septic tank to take waste from the inevitable minimum of three bathrooms. These septic tanks need to be maintained to make sure that proper drainage occurs and that raw sewage is not leaked into the surrounding waterways. This is often not carried out, leading to an overall deterioration in water quality. Many towns, villages and rural areas do not have proper sewage treatment facilities, so water pollution will inevitably worsen and become shockingly visible. This is just one problem among a growing list that is highlighted by An Taisce, the conservation organization that is the thorn in the side of the government at present.

An Taisce, or the National Trust for Ireland, was established in 1948 for the 'Preservation of Places of Interest or Beauty in Ireland'. Some of its founder members were eminent naturalists and statesmen like Robert Lloyd Praeger, Frank Mitchell, Seán McBride and Caerbhall Ó Dálaigh.[8] It received an annual government grant of €70,000 until December 2003.[9] This was just weeks before the about-turn on housing policy occurred. The organization was forced to let two key members of staff go. This attempt to sabotage them has not diluted their resolve to 'soldier on, fighting what it sees as the good fight to uphold the common good rather than the short-term gain of vested interests'.[10] An Taisce stresses that its policies are very much for the rejuvenation of rural areas and not against the wishes of country people. The arguments that it makes for the unsustainability of one-off rural housing are as follows:

Social Reasons
- Increased traffic and car dependence. It is cheaper and better for the environment to create public transport hubs in urban centres.
- As people grow older [and life expectancy increases], they have more need to live near crucial social services like hospitals [which are becoming more centralized after the Hanly Report].
- Only those with access to free sites can build an affordable home in the countryside, and many of the constructions are in fact holiday homes in vulnerable coastal areas.

Economic Reasons
• Increased cost of servicing scattered housing.
• Damage to the tourist industry.

Environmental Reasons
• Groundwater pollution from septic tanks.
• Causing an aesthetic blight on the landscape.[11]

Considering the powerful interests that are explicitly challenged by the trenchant criticisms above, it is obvious that An Taisce will not win any popularity contests in post-Celtic Tiger Ireland. The property developers who would wish to line our coasts with a rash of holiday homes, the politicians who may lose votes on this issue, the builders who do very well from rural construction work, and those who are refused planning permission to build are but some of the infantry lining up across the trenches from this small but vociferous organization. As an example of their outspokenness, Frank Corcoran, chairman, stated:

> It's at the heart of Irish culture to have a whole set of guidelines, policies and regulations to protect the environment, water quality, the landscape, our architectural and archaeological heritage, regional planning and so on, without ever having any intention of complying with them.[12]

They have recently, however, secured a contract from the EU's Environment Directorate to monitor the implementation of EU environmental legislation in Ireland.[13] They are therefore becoming a watchdog body that will have to be taken seriously, whether the government likes it or not.

Critics of An Taisce's policy on rural housing usually make two kinds of argument. Firstly, rural representatives often portray them as being arrogant city people who are only interested in keeping the countryside pristine for the enjoyment of the weekending urban middle class. Secondly, some argue that scattered settlements are more natural in the Irish landscape than nucleated urban ones. They argue that the aspiration towards the development of villages and towns is more British in origin. This argument obviously strikes a chord in Ireland, with its troubled colonial past. The peopled Irish landscape is compared positively with the beautiful but desolate highland estates of Scotland, which are still in the grips of a feudal land ownership structure.[14] Is it not better that the land is owned and settled by thousands of ordinary people than by a small handful of multi-millionaires? This is a somewhat rhetorical question. Indeed, Jim Connolly, spokesman for the Irish Rural Dweller's Association, has said that if people want to see unspoilt scenery, they should go to Scotland instead of Ireland.[15] They argue that it is the physical evidence of human settlement, like megalithic tombs and dry stone walls, that makes the Irish landscape interesting and unique. These critics agree with planning for housing that is more aesthetically harmonious with the surrounding landscape. For example, one commentator bemoans the demise of simple traditional housing styles. Elements of tradition should, according to her, be incorporated into modern design:

> Traditional systems are resources which provide critical bases from which to prise open the cracks of possibility in contemporary consciousness and imagination.[16]

However, as we have already seen, An Taisce go slightly further, challenging head-on some very powerful interest groups.

Whenever An Taisce is mentioned in the media, it usually refers solely to their stance on planning issues. Little or no mention is made of all of the other activities they are involved in, which encourage both children and adults to take better care of their environment.[17] It is interesting to note that their critiques of planning and regional development policy are virtually identical to the recommendations of both major policy documents discussed above. They have not come up with any radical plan that would place them in the political wilderness. They simply serve as a constant nagging reminder to national and local government representatives as to the extent of their new duties. Disgruntled TDs and Councillors use An Taisce as a scapegoat to distract the general public from the unavoidable fact that the more negative aspects of economic wealth must be controlled and fairly distributed over Ireland's beseiged landscape. The environmentalist agenda they represent is relatively unpopular in rural Ireland. Many farmers think that Greens have had too much power in Europe and that they have been primarily responsible for CAP reform and curbing their productivist activities. As already seen in relation to the Nitrates Directive, farmers and their representatives feel under seige from Greens, and it is not appreciated that their influence is also now visible in housing policy. An Taisce now have more political muscle than in the past, however, because they represent the mainstream in EU policy on sustainable development. This would certainly not have been the case prior to the nineties. Such organizations are also highly globalized, in that they use international networks to promote their cause globally.

One organization that is very much interested in maintaining the demographic health of rural Ireland is Rural Resettlement Ireland (RRI).[18] They help to relocate people from urban areas, primarily Dublin, to rural areas, primarily in the west of Ireland. To date, they have relocated 600 families, including 2000 children. Considering the uneven population distribution of Ireland, 'every single family that avails of the Rural Resettlement option will bring benefits not alone to themselves, but also to the country at large'.[19] When interviewed, Jim Connolly, its founder, mused about the detrimental effects on a family of having a 'bad address'. The stigma experienced because of this diminishes with relocation, as they are all viewed equally by locals, and interestingly, he stresses 'there is no such thing as a bad rural address'.[20] He also is proud of the positive effects relocation has upon children's educational aspirations and performance because of improved self-esteem. In their brochure, which paints a very balanced and realistic picture of rural social life, RRI emphasize:

> Small communities welcome positive people who have the drive and enthusiasm to make beneficial changes for the youth, for sport and for community life in general.

As already mentioned, the social composition of localities in rural Ireland is now extremely varied. Even an apparently homogeneous community probably has more social differences than similarities. As already noted, there are far fewer farmers than in the past and a rather low level of farm succession. The children of farmers often go on to third level education to train for a professional career instead of staying on the farm. Despite this, they rarely sever the emotional links with the home place completely. This means that later, when they become established in their career of choice, they may want to return to their locality of origin to build a home for themselves and raise children near their extended families. This is especially the case when property prices are so high – it is a very good start to get a free site from one's parents. It may also be convenient to be near extended family for childcare purposes. Despite the fact that these young people may now have absolutely no contact with farming or the land *per se*, they may feel that they have a greater right to planning permission than a complete outsider. The neighbour's son may be a builder, and it would be convenient for him to build the proposed house rather than having to travel many miles to work every day. Especially if the area is scenic (notably waterfront property), it is inevitable that some outsiders will want to move into the locality, to buy either a first or second home. Whether they are Irish urban-dwellers or non-Irish people, they are often blamed by locals for raising the price of property. Ironically, local people usually want to build rather lavish new homes, whereas outsiders may prefer to buy up old properties and refurbish them to their own taste. In this way, the outsiders may be more interested in local history, landscape conservation, etc. than the so-called locals are. Because it is impossible to determine who is the most deserving in social terms (note the broad terms of Minister Cullen's proposals), it is vital that there are clear transparent processes to determine the most deserving in aesthetic and environmental terms.

Whatever way this goes, the local estate agent is happy to see high prices, those selling the sites are certainly happy to see them, and builders are happy to get the often inflated prices they charge for their services. These separate identities may be contained within the one family, or even perhaps in the one person. There seems to be a rampant hypocrisy in rural Ireland on housing and planning issues, depending upon whether one is selling or buying, or upon who exactly is building the construction in question. This is the case even before we mention anything to do with corruption in local (and national) politics, which everybody knows goes on, but is so difficult to prove. Getting things done the way one wants them depends upon the social resources one possesses. Bourdieu's capitals are relevant here – whether one has the economic assets of land and cash, the cultural assets of education and articulacy, and the symbolic assets of status in the community and connections to the local 'movers and shakers'. The best position is to have all three types of capital, but in some cases, just one may be enough.

This competition is exacerbated in a consumerist culture like Celtic Tiger Ireland, where people want to own the very best of everything (or at least the most expensive), from imported Italian marble bathroom tiles to the latest model of gas-guzzling SUV from the US. It seems to be the case that people are often willing to try to bend the rules in order to get exactly what they want – now. The housing needs of the population obviously need to be met, but it is a question of volume, location and style. A balance needs to be struck between maintaining a demographically and socially vibrant rural population and maintaining the landscape in a way that is beneficial to all. Because of ribbon development, it is even becoming more difficult to find a quiet country road that is not lined with bungalows and in constant use by commuters. The hedgerows that we all grew up with may even be under threat. Will we be visiting heritage centres in twenty years' time, to see what a fuchsia or a hawthorn used to look like?

The untrammelled greed that we now often see needs to be curbed in order to achieve these simultaneous goals of real development and of conservation. While it is unquestionably positive that many more people can now afford the homes and consumer items that give them pleasure, we need to be careful that the distinctiveness of the Irish landscape is not compromised further because of this increased wealth. After all, the land is a finite resource, and the elements that make it so special have already been eroded by inappropriate overdevelopment in some areas. Frank Corcoran, chairman of An Taisce, said recently that the situation in counties Kerry and Donegal 'is so out of control that it can't be stopped ... [because of] schemes being pushed through by councillors against planning advice'.[21] Both of these counties, like many others, are heavily dependent upon the construction sector for male employment. In the wake of several factory closures there, very few alternatives currently exist. This puts all the more pressure on local representatives to look only to short-term gains and push through unsavoury building projects. This is a dangerous downward spiral, which will ultimately lead to the undermining of quality of life in rural areas, as well as a drop in tourist numbers. The ending of such (let us remember, ultimately corrupt) practices, the implementation of a sensible regional development policy and a more imaginative approach to employment creation are all essential in order to foster an economically vibrant society and a rural environment of which we can be proud.

Producing 'Clean' Food

That old phrase 'you are what you eat' is now being taken very seriously indeed by Irish consumers. Concerns over such issues as the genetic modification of food have been rife, and awareness is increasing of the monopolistic position of a handful of global supermarket chains at the retail end of the trade.[22] Five out of ten of the richest people on earth are the heirs to the biggest of these, the Walmart empire, and between them they are worth $100bn. At the same time, their staff work for wages that keep them well below the poverty line and they often work in appalling conditions.[23] The fast-food industry has entered the spotlight too, regarding unsafe and unethical issues in its food production and sale. The rising popularity of such publications as Naomi Klein's *No Logo*, Joanna Blythman's *Shopped* and Eric Schlosser's *Fast Food Nation* highlight this concern. These are but three of the most famous of the new genre of critical publications that are swelling the ranks of the politics sections of our bookstores. The issue of food has become politicized in an unprecedented way. If we look in detail at exactly how our food makes its way from the farm to our plates, it shows itself to be a thoroughly sick industry in which the negative aspects heavily outweigh the positive ones.

To buy organic food, especially when locally produced, delinks one's consumption from this high-risk arena of mass food production. Organic farming uses little or no farm chemicals or animal drugs, builds up soil rather than depletes it, and overall works in harmony with nature rather than seeking to dominate and damage it. For many, buying directly from an organic farmer more than compensates for the higher prices. Some reasons for this are:

- you are contributing to the local economy in a positive way, which helps to counter the market trends that are driving so many farmers off the land in Ireland and all over the world[24]
- the food is produced in the local area, and therefore its transport has not exacerbated the depletion of the world's oil resources, so the number of 'food miles' involved may be measured in single figures rather than in thousands
- there are no unwelcome harmful chemicals in the food that may lead to serious health problems.
- the organic product came from an animal that has been raised in good conditions
- the land on which the food was produced is worked in a sustainable manner, which maintains the integrity of the landscape and respects the wildlife that make their homes there
- you know the people who produced the food

The second half of the nineties saw a huge growth in the number of approved organic farmers in Ireland, at a rate of 45 per cent a year. The number peaked at 1100 in 1999. In 2004 this is now down *c.*10 per cent, to about 1000. The nineties' growth has been attributed to the attractiveness of REPS I payments. The conversion rate then was €349/ha, and after full organic status was achieved, they could get €254/ha. The subsequent reduction in numbers since 1999 was probably caused by the less attractive package in REPS II, of a conversion rate of €180/ha, and after full status, €91/ha.[25] In REPS III (which began in June 2004), the payment rates remained the same as in REPS II, despite the high inflation rate and increasing costs for farmers. Two new amendments were made to REPS III, however, that may encourage more farmers to convert. Firstly, it is now possible to convert just part of one's land to organic, so farmers can experiment with one type of product. Previously, one's entire farm had to be converted all at once. Secondly, farmers can now use commonage for organic farming, having got permission from the other land users. To date, however, very few conventional farmers are willing to convert to organics. If this continues and the new amendments are not enough to encourage them, it could undermine the entire industry. In order to explain this reluctance, John Hoey, chairman of Irish Organic Farmers and Growers Association (IOFGA), muses:

> I believe the problem goes into the perception and notion of Irish people and their history. The past is always seen as a terrible struggle, as the bad old days and anything that isn't modern or clean and spanking new ... is not to be desired in some way.[26]

However, in previous research carried out by this author it was found that there is something of a knowledge vacuum in rural Ireland regarding organics. Several farmers were found to be very open to the idea of converting, but they did not know where to turn to get information and support.[27] These days, however, more people know of at least one place where they can find information: the Organic Centre in Rossinver, Co. Leitrim. This is a centre for information, training and demonstration of organic gardening, growing and farming. It is on a nineteen-acre site, with twelve staff, a demonstration garden, visitors centre and conference centre. This centre fills an information gap for the public and is readily contactable via its website.[28]

The levelling off of the number of organic producers has been attributed to a few interconnected factors, which are difficult to separate and prioritize. Emphasis is placed on one or the other, depending upon which side the speaker is on. The certification of organic produce has become a contentious issue, which has affected how organics is perceived by the state and by the public. An Bord Bia want Irish organic producers to agree on a common national logo, in order to foster consumer confidence and enhance its market share. Eighty per cent of organic farmers are registered with Organic Trust and the other bodies are IOFGA and Demeter Standards (Bio-Dynamics). The split between Organic Trust and IOFGA originated in 1991, when a dispute emerged regarding whether the logo used should specify that the product was also Irish. The dispute was a combination of a disagreement between Irish and non-Irish producers, and between producers and those importing organic goods from other countries.[29] This disagreement continues to this day and is often blamed for the slump currently experienced in Irish organics. If they cannot agree on such an apparently simple matter, the state feels they cannot promote them.[30]

In the short- to medium-term, the market for organic produce is likely to increase, so it would be a missed opportunity if supply could not meet demand. The issue of certification will have to be dealt with by organic producers to improve their credibility, because consumers in the US and UK are already becoming aware of the very different standards operated by the various certification bodies. Irish standards are high in comparison to those of the official UK body, the UK Register of Organic Food Standards (UKROFS), which is one of six certification bodies in the UK.[31] With increasingly open markets in organic produce, low standards of imports thereof could either positively or negatively impact upon Irish organic producers, depending upon how they address the issue.

This question of certification has become a major bugbear. Trevor Sargent, leader of the Green party, asserts that this has become an excuse used by the DAF for its lack of support.[32] Without this support, organics will remain in a marginal position. Tovey argues that the more the organic sector interacts with the state, the more they are turned into 'organobureaucrats', who become removed from the original radical ideals of the organic movement. Organics becomes institutionalized as just another way of marketing Irish food, rather than a radical critique of industrialized food production.[33] She goes on to conclude:

> institutionalisation will profoundly affect the movement while leaving the state relatively untouched – that the organobureau-

crats will become another species of state agents and those who want a 'real alternative' may have to withdraw, regroup and start all over again.[34]

Public attitudes to organics are rather underdeveloped too. Since most Irish people can still see cows grazing in fields, they may feel that regular Irish food is 'clean and green' anyway, without having to pay extra for an organic logo. An educational programme is needed in order to convince the public of the major differences between conventional and organic farming. An Bord Bia report that the number of people who were willing to pay an extra 10 per cent for organic produce dropped back to 44 per cent of consumers from 57 per cent in 2000.[35] However, in order to address this matter, the 'regrouping' suggested by Tovey in 1999 may already have occurred, in the form of farmers' markets. These have the advantage of 'cutting out the middleman', where producers can do their own marketing and distribution, selling directly to the public. This localization of organic food has created a mini food revolution in many parts of Ireland.[36] This will be discussed further in chapter seven. The prices charged for organic products in the supermarket are sometimes off-putting for the consumer, so direct sales at farmers' markets are one way around this problem. In that way, producers can charge less and still make a profit, and consumers can afford to buy more organic food. It has to be said too that no official statistics reflect the amount of organic goods sold at these markets, and perhaps more importantly, the amount of trust and goodwill generated in this context.[37] This is also allied to the growth of the Slow Food movement in Ireland, as many of these 'niche producers' are members thereof. This is an international movement spanning 70 countries and with 60,000 members. It originated in Italy, and is committed to producing and selling organic, seasonal and locally produced food. It is about reconnecting people with how their food is produced. It currently has about 350 members in Ireland.[38]

However, these markets still do not fully solve the problem of organic food marketing on a grand scale and the resulting status of organics in Ireland. Even in 2004, it is still not taken seriously as a real alternative by the state. Trevor Sargent argues:

> Other countries in the EU view organics as the next stage in sustainable farming, whereas our Government labours under the notion that the organic sector is ... slightly exotic, like ostrich farming. It's a ridiculous state of affairs.[39]

How does one explain this attitude? One could suggest that An Bord Bia may be reluctant to put too much emphasis on this sector, as it may invite negative comparisons and hence cast aspersions on conventional Irish farming. However, without this support, organic producers are left out in the cold, to brace themselves against the hurricane of the open market. This Catch 22 situation will have to be addressed head-on to prevent organics from becoming even more marginalized than it already is.

A problem has emerged in England recently in relation to organics: fraud. Spot checks by Environmental Health Officers have been conducted on some shopkeepers and market stallholders and some have been found to be charging premium rates for food that is not organic at all. The *Observer* newspaper carried an exposé on 21 August 2005, claiming that 'sharp practices are widespread' in British organic farming, like spraying crops and vegetables with chemicals and non-organic manure. While there is absolutely no evidence of such carry-on in Ireland, it is important that organic producers keep a constant eye on such developments and that the trusting relationships between consumers and their organic suppliers are maintained.

In the overall European context, the future of organic farming is challenged by the likelihood of more controls being placed over the amounts of inorganic fertilizer used on crops. With an overall improvement in standards of food production organic produce would no longer need to fill this niche.[40] The Single Farm Payment that farmers receive from January 2005 will depend upon adherence to several EU Directives, the most important of which is the Nitrates Directive. This could improve the reputation of conventional farming in the long run, hence organic producers need to take this into account when marketing their goods, with or without state support.

6. Contrasting Rural Visions: Two Key Issues

Caroline and Eddie Robinson grow vegetables without chemicals in Co. Cork. Even though they operate a highly ethical thirty-acre farm that adheres to the highest organic standards, they term their produce 'chemical-free' rather than organic. They sell their produce directly to customers, most of whom have visited their farm on 'open days'. In that way, they do not need to have a Symbol and so they escape what they view as a restrictive bureaucratic framework imposed by the certification bodies of Organic Trust or IOFGA. Caroline is the Chairperson of the newly formed Irish Food Market Traders Association. This group is currently engaged in negotiations with the Food Safety Authority of Ireland, who have been sending Environmental Health Officers (EHOs) round to markets and small food producers and making their lives difficult by demanding that they attain unrealistic hygiene standards. She is very committed to farmers' markets which she considers to be 'a lifeline for rural towns, because the money generated stays in the area'. When asked why she thought organic production had levelled off in recent years, she highlighted the fact that the bottom fell out of the British market for Irish organic beef after the Foot and Mouth outbreak. Following this, Britain established a marketing tool called the 'Red Tractor symbol', which is designed to encourage the public to buy only British produce. However, regardless of the macro-scale, she is optimistic about the future for small food producers, because she sees so much demand for 'real food' that is low in hydrogenated fats, excess sugars, salt, preservatives and chemical additives. She eagerly awaits new EU legislation which is supposed to be more lenient on farmers' markets, following lobbying from the Slow Food movement. She sees no reason why An Bord Bia would not approve of this, as large food producers like the dairy co-ops can benefit from the gains made by the small producers. If Ireland develops a reputation for excellent food, it benefits everybody in the long run. It can also benefit the health of the nation, which is plagued by so many problems like obesity and diabetes, which are directly related to poor eating habits.

One of the better-known brands in Irish organics is that of Ballybrado, from Cahir, Co. Tipperary.[41] Many different types of goods are sold under this label, from biscuits to chickens. The owners of this company have now undertaken two major new initiatives. The first is an organic meat factory, where organically produced animals are killed and packaged for sale. Secondly, they have now used this location as a base for a 'virtual organic farmers' market'. This is a distribution system, where customers can shop online at their website and have wonderful organic food delivered to their homes. This cuts out supermarkets and distributors, meaning that the goods can be sold at a substantially lower cost. With innovative ideas like this, the future looks bright for Irish organics.

A related issue here is the ethical one of animal welfare. Organic farmers generally have higher standards of animal welfare because the animals are given straw bedding and afforded more space to move about and lie down than in conventional farms, in which slatted houses are the norm. A recent study conducted by Teagasc on animal welfare came to the same conclusion.[42] In this study, they tested the 'Animal Welfare Index', which is used in other European countries like Austria, Germany, Netherlands and Switzerland. The criteria that comprise this index are opportunities for movement, eating/drinking facilities, opportunities for social behaviour, resting facilities, comfort and exploration facilities, hygiene and animal care. This comparative study researched these on both conventional and organic farms and the latter scored much higher on most of them. One German organic sheep farmer interviewed by this author described his *modus operandi*:

> Whenever somebody comes from the authorities, from organics or REPS, they are very surprised at what we have here ... We deliver our organic lamb up to Newcastle West to the organic factory, it's a long way, but I'm going by the Alpine laws, where animals are not allowed to be transported any longer than four hours for slaughter, breeding animals are allowed to be transported for 2 hours a day. I can afford the luxury to play around a bit ... I'm not a fully-fledged farmer, so they are all more or less like pets to me, even when we had the 350 sheep.

Treating the farm animals 'like pets' is a sharp contrast to the housing conditions of farm animals on most of the larger conventional farms. On smaller farms, however, farmers may often have a surprisingly close relationship with their cattle. One farmer interviewed here described how he felt about his animals:

> We used to milk forty-five cows here but in 1983 we were cleaned out with TB. We had a grand herd of cows and ... the only time I cried was when I saw my cattle going ... all the cows were in calf

and to see two lorries of them leaving that yard, and milk dropping out of them. I did cry, it put me back years.

The most common set-up employed by farmers these days is
the slatted house. This involves flooring with slits where the
manure can flow down into piped channels towards the slurry
tank, from where it can be safely removed at the appropriate
time. This is the optimum system, encouraged by Teagasc farm
advisors, which manages slurry in the most efficient manner
and is most convenient for the farmer. From this perspective, it
is highly efficient, but from an animal welfare perspective, it
leaves a lot to be desired. Farmers interviewed here reported
that diseases spread faster among the animals in these conditions, and they often got foot-rot and sore feet from standing
on slats all day long. One farmer said:

> I don't think that slatted houses are the places for an animal to be
> lying on top of, with a load of gas coming up. Animals during the
> winter should have a certain amount of natural light, it's been
> proven that animals do much better that way. The reason that slat-
> ted houses proliferate in this part of the world is convenience and
> also straw bedding is not available, in Denmark, it's all straw bed-
> ding because they've got loads of it, so they don't build slatted
> houses there anymore.

It is completely inevitable that concerns such as these will
become part of the official lexicon of the EU within just a few
years. It is certain at least that three Directives will form part
of the cross compliance regulations for receipt of the Single
Farm Payment in January 2007. These concern standards for
the treatment of all farm animals and especially calves and pigs.
Lots of farmers will be in serious trouble when these requirements come into play. Some Teagasc researchers advocate:

> The Irish beef and dairy industries are our most important agricul-
> tural assets. Pride in the welfare practices employed in a 'clean and
> green' system of production will undoubtedly inspire consumer
> confidence in the food (beef and dairy products) we produce.[43]

Teagasc itself seems to be on a learning curve on this issue,
because while their foot-soldiers have strongly advocated the
construction of slatted houses over the years, their researchers

are only beginning to recognize the economic (and perhaps even moral) merits of kinder housing conditions where the animals can move around and lie down on straw bedding.

A subset of this issue is the issue of animal transport. Live animal transport forms a substantial proportion of agricultural exports from Ireland. The conditions under which the animals were transported were deemed to be a problem, so the rules were changed in 2003 by the EU Commission. These stipulated:

- a maximum of nine hours travelling and minimum twelve hours rest
- no staging point is required: animals should have feed, drink and rest in the vehicle
- increased space allowances
- permanent access to water
- young animals not allowed to travel more than 100 kms
- a ban on transport of females one week before/after giving birth[44]

Depending on one's perspective, these new rules seem either over-restrictive or not radical enough. Farming interests are very unhappy about them. One spokesman said:

> the animal rights activists [the Commission] are listening to will not be satisfied until Europe converts to Hinduism and makes cows sacred. As for farmers and processors, they can play the role of the Untouchables.[45]

On the other hand, the main animal rights organization in Ireland, Compassion in World Farming (CIWF), said that they were very disappointed because the new rules did not go far enough. They wanted to see an overall journey limit being imposed, meaning that carriers could not repeat the above travel pattern over and over again, up to perhaps ninety hours of travel.[46] Despite these diverging views at either side of the ideological spectrum, the above rules have now become institutionalized, and a permanent feature of the current political climate. It is but another warning to the farming sector, for better or worse, that the old practices they they previously took for granted will be challenged more and more in future. This may serve as an encouragement to some to diversify into more sustainable enterprises that meet widespread consumer approval, but no doubt for others, the only challenge presented will be how to bend or break the rules. There is risk involved in either option, and it depends on where one stands on the social ladder as to which is the most likely choice made.

Conclusion

The two issues highlighted for discussion in this chapter illustrate the difficulties involved in taking the ideal of sustainable development beyond mere rhetoric and into the realm of reality. Broad guidelines have been issued on both issues by the EU. Rural housing is to be based on need, and is not to compromise the environment. Organic farming is to form a central plank of this current phase in agriculture, where productivism cannot be tolerated as widely as before. The state's input leaves a lot to be desired on both issues. Local politicians often undercut planning regulations on housing and there seems to be little commitment to regulated planning at Cabinet level. Organic farming does not receive the support it needs from government agencies and the sector seems to be mired in an argument about certification. Interest groups and very dynamic individuals exist in relation to both issues, expressing varying opinions and trying to move the debates along. These people and interest groups are allied in various ways to international movements. Above all, the general public muddles along regarding both of these issues, trying to make the best of their situations, regardless of the 'big picture'. They are both microcosms of social life in general, in which conflict is inherent, and how one fares in the negotiation process depends upon how one uses ones social resources.

chapter 7

West Cork: Competing Views of a Region

Having discussed general farming and environmental policy developments, it is now time to isolate a rural region and to attempt to analyse the social dynamics that shape its character. west Cork is a region of Ireland that is close to many people's hearts. There are a number of ways in which it may be viewed, from a cosy rural getaway to a place of economic decline. It is a matter of perspective and all are correct simultaneously because the area is undergoing rapid socio-economic change. In general, farming is on the wane in this area, being replaced by tourist-related activities. This is changing the social character of the place in numerous ways.

The social actors in a place like west Cork may represent interests, or networks of interests, that stretch far beyond the physical boundaries. While the objective facts that create the socio-economic and geographic composition of such a place need to be mapped out, a mental map will no doubt be made up of much more dispersed and dislocated co-ordinates than those we find in an atlas. Each individual is the product of any combination of varied cultural, economic, social and political perspectives. Hence any spatial area, being comprised of complex and varied physical realities and different sets of individuals, can produce varied results, or what we may term a meaning-cluster. Because of inequality between social actors, certain types of meanings become dominant and others subordinate at any particular time. Applying this idea to west Cork, of what ideas and constructions can we say that it is comprised? What type of meeting place is it, and which types of socio-economic relations shape its development?

A Region of the Mind?

Visiting the town of Bantry on the first Friday of each month, one will find the square in the centre of the town taken over by a colourful and eclectic market. I usually visit this market, as did my father when he was a young man back in the forties and fifties. He went there for different reasons, however. He was a cattle dealer who toured all the cattle fairs to chat, haggle and ultimately to buy cattle that were transported by train on the West Cork Railway. Bantry was then at the heart of a poor but lively rural landscape that was driven primarily by fishing and subsistence agriculture. Both the monthly cattle fairs and the local railway met their demise in the late fifties and early sixties at the beginning of the state's modernization drive, which was intended to save rural Ireland from itself.

The square in Bantry no longer resembles the one in which my father bought and sold cattle. It is now paved with expensive paving stones, with public seating and a central fountain, paid for with EU aid money. The statue of St Brendan that presides over this scene was donated by Gulf Oil in the seventies, then a major employer in the town. This was before the Betelgeuse disaster of 1979, in which over fifty people lost their lives and the whole harbour was devastatingly polluted with crude oil.[1] The recently revived fair now attracts a much more heterogenous clientele than the hawkers and dealers from the agricultural hinterland of my father's time. Now, alongside the farm chemicals and wellington boots are stalls selling healing crystals, second-hand clothes, Indian tie-dye throws, Mexican jewellery, olives, dodgy antiques, goat's cheese and organic vegetables. The vendors of these goods have come from far and wide, but they all have one thing in common. They have moved to west Cork in search of a sense of community and meaning, having bowed out of their previous lifestyles. They are more middle-class than lower, more cosmopolitan than local, more socially diverse than homogenous. The combination of all these people's dreams has created a social collage that is epitomized in the new Bantry Fair, an indicator of just how much has changed in this area of rural Ireland since the fifties. In the words of the much-lamented Pete McCarthy:

> Today, it's a glamourous destination, a haven for upmarket tourists, English expats, and Dutch cannabis importers, but in the 1950s and 1960s it was the arse end of the back of beyond, and that may be talking it up.[2]

The west Cork of the struggle for independence that produced renowned nationalist heroic figures like Michael Collins and Tom Barry no longer has the resonance that it once did. This meaning-cluster, based on nationalist ideology, is now dominated in popular discourse by the idea of the postmodern, multicultural west Cork, with its world-renowned bird observatory on Cape Clear island, its independent music festivals and Babel-like pubs. Some parts have enormous *caché*: a quarter-acre site recently went on sale outside Kinsale for €650,000. But in what sense is it a region *per se*? It is one of those regions for which it is difficult to establish either an analytical or physical border. No region can be analysed in isolation, and social theory needs to work in tandem with geography to create an analysis that

> treats regions as dynamic components of an unfolding world economy, rather than as closed worlds or discrete components of national and global entities.[3]

This region is also physically ambiguous. A report drawn up by Teagasc in 1994 took it to be comprised of the Rural and Urban Districts of Bantry, Castletown, Clonakilty, Dunmanway, Skibbereen and Schull.[4] They found the population of this area to be 47,608 in 1991, with 40 per cent working in the census category of 'agriculture, forestry and fishing'. The towns of Bandon,

Macroom and Kinsale are at the edge of the area. An alternative delineation corresponds to the Cork South-West electoral area, which incorporates Bandon and Enniskeane. This region has its own radio station and its own Enterprise Board.

Definitional difficulties emerge, depending on whether we are concerned with adhering to rigid bureaucratic boundaries, or whether political and cultural associations are more important to the analysis. Massey argues:

> 'Boundaries' ... are not necessary for the conceptualisation of a place itself. Definition in this sense does not have to be through simple counter position to the outside; it can come, in part, precisely through the particularity of linkage to that 'outside' which is therefore part of what constitutes that place.[5]

The latter option offers more analytical potential because west Cork is a space that is a popular idea, a form of identity; the delineation exists largely in people's minds rather than as a distinct region. Several of these type of regions exist in Ireland, from Inishowen to the Burren to the Beara Peninsula. While these feature strongly in everyday life, they rarely correspond with administrative boundaries. Kevin Whelan reminds us that these regions:

> offer a humane and comfortable sense of scale and attachment which is not imposed from without but arises from within a community's own sense of shared identity.[6]

Far from being a closed, isolated or peripheral region, west Cork has a long history of trade and cultural connections with the wider world. Thus, for example, its richer areas were given to the production of agricultural commodities for the export trade, including trade with Britain, France and Spain and with the western Atlantic colonies. It imported wine from the Mediterranean world, with the small ports and coves of the region also functioning as safe havens for pirates in the fifteenth and sixteenth centuries. The coastal area also had a thriving fishing industry throughout much of the nineteenth and early twentieth century. Much of its regional autonomy was curtailed, however, with the development of British state institutions here, including coastguard stations and police barracks, which were constructed from the 1820s onwards. The late 1830s and early 1840s saw the further elaboration of state control over the region with the construction of workhouses in the area, most notably at Skibbereen, Bantry, Clonakilty, Ballydehob and Dunmanway.[7] It was from this period onwards that many local peasant agricultural practices were denigrated, as the rationalization of agriculture and land privatization proceeded apace.[8]

West Cork is the meeting place between interlocking worldviews, some of which are complementary, and others conflicting. The relations that occur between actors with different amounts of economic, cultural and symbolic capital produces varied results in the area. Localities are created out of these intersections and interactions of concrete social relations and processes. Rural areas such as these are subject to competing constructions, as they can now represent a greenfield site for industry, a place of rest and recreation for the urban-dweller, a conservation site or a place for farmers to make a living.

West Cork is very popular for both tourism and in-migration, and this has ensured that it is at the epicentre of the current property boom that has gripped Ireland. Waterfront property is at a premium and the market value of previously undesirable houses has now reached unprecedented levels. Local farmers cannot help but be tempted by the extremely high prices that can be obtained for sites for housing development, despite their historical aversion to letting go of the land for such purposes. The result is that some areas, especially overlooking the coast (such as Baltimore), are now very congested, with the small winding roads that surround and intersect it completely unable to cope with the volume of traffic that it attracts, especially at peak tourist season from May to September. The pressure this congestion places on the local infrastructure has yet to be addressed as a serious issue. As yet, too much money is being made from tourism for any challenge, or even control to be placed upon this sector. Local heritage sites are affected because of all the construction that is going on beside historically valuable old walls and buildings.[9] An Taisce

West Cork have issued a handbook advocating the retention of 'traditional' housing styles that blend into the landscape, in an attempt to contribute positively and practically to debates about rural housing.

Competition between different social groups in such contexts can often turn into conflict. An example of this is the generally negative perception of New Age Travellers, especially by local business interests. This group is generally rich in cultural capital, but chooses to express it in a way that challenges the established mores of Western society, i.e. upward social mobility and consumerism.

An example of a more overt clash of interests occurred in recent years between three groups of social actors in Bantry: traditional fishermen, the new mussel farmers and those involved in the tourist industry locally. Since the Whiddy Oil disaster in 1979, there was a huge community effort to clean up Bantry Bay. This led to the development of some mussel farming in the bay and in nearby Dunmanus Bay, to generate alternative employment. This is now well established, employing 150 people. However, recent attempts by another local businessman to expand this into neighbouring Dunmanus Bay has created new problems. It interferes with traditional fishermen earning a living in the inlet – to which they feel they have a 'historic right' – because it is so difficult to manoeuvre their boats between the mussel-lines. Local people who depend on tourism for a living also object because they view the distinctive blue barrels that mark the mussel-lines in the water to be highly unsightly and off-putting for tourists, 'a blot on the seascape'.[10] This local conflict remains unresolved.

A more recent conflict takes us further west to Castletownbere. The film director Neil Jordan, who owns a lavish home near the town, has blocked the construction of a large holiday home and leisure development there. Jordan and one other intermittent visitor from Dublin were the only objectors to the plan for forty-one holiday homes, a clubhouse and swimming pool. Many locals are resentful of this, as they felt it would have created much-needed employment in the area.[11] Naturally, however, it is mainly those who object whose voices are heard on the matter.

It is therefore a fruitful approach to examine the various competing constructions of the area. West Cork, like any other place, is several different places at once. It can be painted in dark or bright colours, a place of light or shadows.

People Coming and Going

West Cork is an area of dynamic demographic ebbs and flows, thus the region has affinities with many other places on the globe: the destinations of emigrants and the derivations of immigrants. Emigration is a well established fact of life in the region since at least the nineteenth century, immortalized in such laments as 'Old Skibbereen'. Commins and Frawley (1994) used census data to establish how many young women and men were 'missing', by comparing the number of age cohorts of young adults (9–25) with the number of cohorts when they were 5–9 years old. In this way, they found that 4 out of every 10 young adults were missing from the region, with population decreasing in rural areas and increasing in the towns. This is quite a crude measure of emigration and serves to remind us that census data alone is rarely adequate to measure social phenomena. While charting the trend in a basic manner, it does not provide any qualitative information on the type of out-migration that occurred, where they are living, their quality of life, and the feelings of those they left behind.

MacLaughlin (1994) adopted this approach in a relatively recent national survey on emigration, part of which was focused on west Cork. Of the 166 families surveyed in Skibbereen and Schull/Ballydehob, 60 per cent had one member abroad, and the remaining 40 per cent had two or more. MacLaughlin terms these 'Ireland's multinational families'. The majority of emigrants from west Cork at that time were still going to the traditional destinations of London and New York, had relatively low educational levels, and were working in relatively unskilled jobs: construction for the men, services for the women. Most poignantly of all, many returned to visit home more than twice a year. This was at a time when, to legitimise the veritable exodus that was occurring, the government rhetoric said that emigration had nothing but a positive impact and gave young people new skills that they could use on their inevitable return home. An updated study would fill in the gaps as to whether patterns of emigration have changed since the late eighties. With the information available now, we can at least suggest that places like the Bronx and Kilburn, in which there are large clusters of Irish emigrants, are very much part of west Cork's extended socio-economic and cultural world, and co-ordinates on west Cork peoples' mental maps.

West Cork has become a major focus for counter-urbanization.[12] It has a reputation for a relatively high level of in-migration, in the form of both Irish and foreign new residents, second-home owners, European retirees and New Age Travellers. These incomers are sometimes derisively termed 'blow-ins' by local people, 'as a device for class levelling'.[13] They have chosen the more robust lifestyle of the west Cork countryside, eschewing the convenience of urban life. One Irish artist who has been living in the area for several years describes the immigrant women and men in the following fashion:

> The women of the immigrant group tend towards the virtues of strong character and great capacity: bearing children, baking bread, turning their hands to dairying, craftwork – any task demanded by frontier life. The menfolk are more purely decorative: handy for rolling a joint, fathering a child, claiming welfare.[14]

Very little is known about these diverse inhabitants of west Cork in sociological terms, but even the most cursory drive through the area demonstrates a significant level of cultural diversity, uncommon enough to Irish rural areas. The fact that sauerkraut and German sausage is sold in most small west Cork supermarkets serves only to hint at this. The most common type of housing they prefer is an old cottage or farmhouse to renovate, stripping the plaster off the external walls, exposing the bare stone. These are usually complemented by window-frames painted in bright colours (usually red) and lots of flowers in the inevitably impeccable surrounding garden. While conducting this research, one elderly farmer said to me:

> People used to look down on the people who lived in cottages, but the best of people are moving into them now!

The incomers are both counter-urban and counter-cultural. They are not deeply rooted in the area, and many survive on the income from small businesses. They often live in nationality clusters (like the English in Skibbereen), they live mostly on the coast, and they live according to separate socio-cultural geographies. It is probably the case that this group derive from the middle class and upper middle class in their places of origin. They have a more conscious sense of cultural association, since they have no family history in the area upon which to rely for identity. They are self-contradictory in a way, because while they may be cosmopolitan in their outlook, they are still searching for an organic community in the area.

Another study also found that the migrants to a remote rural community in the north of Scotland were mostly of a middle-class or upper-middle-class background. This group, termed 'urban refugees', 'do not come to the countryside to practice their previous professions but to make a radical change in their own style of life'.[15] Also, length of residency can lead to some internal differences within this group. This was also found to be important in a study of a village in the Yorkshire Dales.[16] There, as here, there is a distinction made between those incomers who had lived in the area, say, since the sixties – the 'old incomers' – or recent arrivals, the 'new incomers'. Some of the people who had been living in the area for fifteen years or more bear some resentment towards those incomers who were now building second homes, and who are not seen as committed to contributing to the long-term future of the area.

West Cork has also become a haven for mostly foreign-born artists, musicians and craftspeople, leading to the growth of what we might term a 'creative community'. The region's physical beauty has inspired artists for some decades now, but especially since the sixties. An artist's co-operative was set up in

Ballydehob at that time, and many other group initiatives have surfaced since then. Imaginative individuals sought a less stressful environment than that provided by the materialistic cities of the core of European capitalism. The attitude of the locals to these 'blow-ins' ranges along a spectrum from total inclusion to total exclusion. This former view could be based upon a genuine openness to and interest in other cultures. Conversely, it is possible that they simply welcome the high prices that they are willing to pay for property, and often for old houses that locals would not usually even consider buying. It is also possible that they appreciate that the presence of the 'blow-ins' may raise the public profile and social status of the area. In general, those who have a higher social status are more welcome. There is an unspoken blanket rule, however, that the 'blow-ins' are expected to strive to adapt to become more like the locals, and their efforts in this regard are rewarded with social acceptance. This was also found by Peace in 'Clontarf', an Irish village:

> Some of these ['blow-ins'], even after a decade or more are considered, and consider themselves to be, marginal to the mainstream of community life: others within a shorter period have become actively involved in that mainstream.[17]

An interesting example exists in Skibbereen, where a famous English actor, Jeremy Irons (who is married to Sinead Cusack, an Irish actress), bought and renovated a local castle. He is very popular and is often asked to participate in the public life of the town, such as opening the annual summer festival. The locals appreciate his obvious love of the area and his ability to be normal and down-to-earth, blending in without being pretentious.

Outsiders who move into rural areas have the advantage of more social freedom, but the other side of the coin is less social acceptance and cohesion. Often their only recourse is to lower their expectations and/or rely on each other for friendship and support. They are not connected to the place in an organic manner, hence they can have a freer rein to make their voices heard and engage in public-oriented activities. If littering, for example, is a problem, they are less likely to keep quiet about it than locals. Anecdotally, it has been known for locals to put a 'word in the ear' of one of these 'operators' because it lets them off the hook and yet they know something will be done about the problem in question.

Local Economic Development Processes

An undoubted contributory factor in the high rate of emigration has been the decline in agriculture in the region. Commins and Frawley's report on the west Cork region in 1994 found farming to be waning in economic importance. 39 per cent of its population worked in this sector in 1986, compared to 52 per cent in 1971. The trends within farming are: a) a decline in tillage area of 80 per cent in thirty years, b) a decline in dairy cow numbers of 20 per cent in the eighties alone, c) an increase in the numbers of suckler cows and calves, and d) a substantial increase in the number of sheep, by 130 per cent in the eighties alone. This would indicate a significant move to direct payments as a form of income, like the headage payments for sheep. As in the rest of the country, there is a high level of dependence among smallholders on non-farm income earned by the farm operator and/or spouse, on direct payments and pensions, and on letting land. Teagasc's objectives for the agricultural sector in this region are 'maintaining competitiveness, creating wealth and value-added, achieving more productive use of resources and maintaining the incomes of those displaced from commercial activity'. Whether all of these aims can be achieved simultaneously is very much open to debate.

The alternatives to conventional agriculture that they suggest are forestry and fish-farming, sectors over which doubt has been cast regarding their social and environmental impact. By the late nineties, around 1100 farmers had joined REPS in west Cork, i.e. about one-fifth of total farmers. The relative poverty of the farming land and the small farm size means that farmers here are heavily dependent on EU remittances and part-time farming. The extensive nature of farming in the area also means that very few changes would probably have to be made by farmers to qualify for an agri-environmental scheme such as REPS. However, the type of farming here does not appear to be quite as poor as some other areas of the west of Ireland or the Border counties, where almost all qualified farmers would have joined REPS.

As well as being an area of agricultural decline, nor did it ever have a strong industrial base. Commins and Frawley (1994) show that in 1991, the area contained only nine manufacturing plants employing more than fifty people, of which 4 were dairy co-ops, 2 timber plants and 2 multinational branch plants. Together, these plants employed only 1000 people. Considering the pressure that has been placed upon dairy co-ops by large retail outlets like Lidl and ALDI, the future prospects for this type of employment are worrying indeed. Many small enterprises exist, but they need to be highly innovative in order to survive the savage competition of the open market. The next section looks at one attempt to encourage such innovation.

Revitalizing the Local Economy I: LEADER and Fuchsia Brands
West Cork is very consciously adapting and re-inventing itself, by establishing itself as an important tourist centre. Due to the relative stagnancy of other economic sectors, there is increasingly heavy dependence on tourism as a vehicle for job creation. As throughout the rest of rural Ireland in the past decade, LEADER has been very active in west Cork in attempting to initiate 'bottom-up' development by building entrepreneurial links and encouraging economic investment along these lines. LEADER 1 ran from 1991–4, LEADER 2 from 1994–9, and LEADER+ began in 2000 and is scheduled to run until 2006. Serious efforts have been made 'to promote locally-based collective action and to integrate rural development efforts across sectors'.[18] It has been seen as a means of encouraging local initiative and reducing dependence on the state. The measures under which LEADER funding is granted are:

- technical assistance
- training/recruitment assistance
- rural tourism
- small firms/craft enterprises
- agricultural, forestry and fishing products
- preservation of heritage and environment

These concur with the main direction in rural development policy in contemporary Ireland, that of transforming land and people that were previously engaged in agriculture to other newer roles, mainly in tourism and rural recreation. This is summed up in the following executive summary of west Cork LEADER Co-Op:

> The appeal of the area should be developed around the environment and eco-friendly practices; activity and special-interest holidays, local produce and cuisine; small-scale low-density tourist product; and improved access through Cork ports.[19]

An initiative has been launched by LEADER in recent years that aims to support the small and medium-sized businesses in the area. One of the most popular images of the region is that of beautiful scenery and a clean and unspoilt environment. This is now being capitalized upon by Fuchsia Brands Ltd., which markets west Cork tourist enterprises and food products under its own indigenous brand. Both of these sectors could be seen as dependent on the regional identity, and a third sector, craft production, is soon to be added. The logo used, an image of the red and purple fuchsia flower is intended to be reminiscent of small, pretty country roads in which the fuchsia bushes are flourishing. The fuchsia logo takes two forms – for tourism, it is surrounded by the words 'A Place Apart – West Cork', and for food 'A Taste of West Cork'. Initiated in February 1998, this company is comprised of representatives from West Cork LEADER, West Cork Food Producers Association and Cork/Kerry Tourism, and was developed by a combination of inputs by EU funding, food economists and tourism interests. The EU provides only 50 per cent of funding for the initiative, based as it is on the principle of matched funding. The project is backed up by significant investment by selected private companies, especially in marketing and specially tailored quality training programmes. The 1997/8 annual report highlighted the facilitation of Ireland's first regional brand as the primary achievement of the year.[20] It received almost half of all funding from LEADER II in west Cork. According to its website, its principal objectives are to:

- harness the distinctive image of the region as an aid to competitive advantage
- promote the region with emphasis on the unique environmental, cultural and heritage resources
- develop the branded identity for local goods and services using local natural resources to satisfy specific market needs
- integrate the development and marketing of complementary economic sectors
- achieve greater collective action by public and private sector local development
- use regional profile imagery and identity to stimulate enterprise development[21]

The CEO of Fuchsia Brands asserts that it is an attempt to gain competitive advantage by 'turning around peripherality'.[22] He also emphasizes that membership of Fuchsia Brands is an important networking strategy for small and medium enterprises (SMEs). They can gain access to a 'broader support structure, help in terms of information and liaison with retailers'. On the one hand, the ideal is that a clustering occurs, where collective learning can take place. On the other hand, he suggested that some SMEs might feel that they had no choice but to join because they might feel that their branded competitors might have an edge that they did not. Using this strategy, west Cork is marketed by LEADER as an entrepreneurial seedbed and a potentially prime site for new investment. At this point, let us briefly discuss the two main 'branches' of the Fuchsia initiative: tourism and food.

In the multi-functional countryside, the ideal is that farmers are reassigned, instead of farming, to producing landscape for the pleasure of urban consumers. The self-same agricultural policy that was largely responsible for substantial rural depopulation on the one hand is now repackaging it in a tourist-friendly, sanitized version of itself that is frozen in time. This means that 'the local sense of place is being replaced by an out-sider's view of what is significant in the locality, i.e. the out-sider's sense of place'.[23] This development approach may also be found to exclude those local inhabitants who are outside of the entrepreneurial milieu and the class groupings that foster this ethos.

Small tourist enterprises have developed all over the area, ranging from self-catering cottages to pitch-and-putt courses to French and Japanese restaurants. An estimated €80 million enters the area every year in tourist revenue, and major plans are afoot to provide even more tourist amenities, like facilities for outdoor and cultural pursuits and more hotel bedrooms.[24] It is estimated that the equivalent of 1200 full-time jobs are dependent on tourism here, and jobs have been created in extremely peripheral areas where little or no other employment exists. Of the six funding categories listed above, the largest amount of funding went to rural tourism under LEADER II, a total sum of just over €2 million. This compares, for example, to €255,000 for Preservation of Heritage and Environment, and c.€140,000 for the Small Firms/Craft Enterprises category. Enterprises supported in the rural tourism category included €63,500 for a tropical visitor garden, €63,500 for an exhibition garden, €63,500 for a sailing club, €31,750 for a diving centre, €63,500 for a golf school, €22,850 for a marina development, €21,600 for the purchase of a steam engine by a vintage club, €8890 for 'river development' by an angler's club, and €6350 for upgrading a bed and breakfast. West Cork LEADER claim that 89 full-time jobs and 70 part-time jobs were created during the LEADER II period.[25]

While tourism does create much-needed employment, it is vulnerable to a number of difficulties, such as a high level of seasonality, labour shortages, geographical distance from the Dublin area and increasing competition from other regions. The jobs created are often low-grade, low-paid, seasonal and unreliable in character.[26] The treatment of immigrant workers in the tourism sector has not yet been properly researched, but one suspects that it will become more of a public issue in years to come.

The second developmental thrust that (literally) feeds both the local and tourist markets is that of high-quality, niche-market artisanal food products. These products range from cheeses like Durrus, Gabriel, Gubbeen and Milleens to yoghurt, milk, butter, eggs, honey, jams, herbs, smoked fish, shellfish, cakes and desserts – all with locally sourced ingredients. These enterprises are sometimes owned by locals and sometimes by incomers to the area. The latter have made a major contribution to the development of a local food culture. The Fuchsia Brands CEO feels that the multiculturalism of this sector is an asset to be celebrated and encouraged, and part of the distinctiveness of west Cork. These entrepreneurial incomers are viewed as contributing to the local economy by providing employment. Indeed, many of the gourmet products are quite exotic in Irish terms, and could by no means be described as native to the region or part of a distinctive regional cuisine. High quality cheese was never part of the Irish diet until it was more or less introduced by German, Dutch and English incomers in the past twenty years or so. Indeed, much innovation is in evidence, embodied in products such as chowder, salsa, duck and cheesecake.

In the light of the enormous power of multiple retailers, one producer said that it was 'a tool to get our products on the shelf.'[27] The advantages of networking and joint marketing under a logo associated with quality were also apparent to the members. The attitude of small producers to the Fuchsia initiative varies from those who rejected it as 'an exercise run by men in suits', to those who gladly took the opportunity to upgrade their operations with regard to hygiene requirements.[28] This suggests that small food producers are caught in a bind, that they perhaps have to sacrifice some of the independent spirit that characterizes their very lifestyles in order to upgrade production standards set by other more powerful actors in the EU and the retail sector. Additionally, the high proportion of incomers who are responsible for developing quality foods and services here led Ilbury and colleagues (2001) to the worrying conclusion that 'the level of entrepreneurship in the native and residual population is not sufficient to fully exploit the natural resources of these areas, especially through adding value'.[29]

This is an example of a type of economic development that on the one hand builds upon a local resource base and labour market, and on the other, is geared towards the international consumer. New concerns about food safety and the increasing sophistication of the tastebuds of a large portion of the population of the west has led to the creation of a market for many new products. The strong symbolic power that Fuchsia products possess ensures that they have a good chance of success at a time when the issues of the homogenization of food products and declining standards in their production have become important public issues. This capitalizing on the idea of authenticity is an opportunistic, postmodern marketing strategy that is often associated with Italy, and most famously with Parmesan cheese.[30] It is now in entrepreneurs' interests to re-value such local characteristics because EU policy-makers now recognize the high consumer demand for the non-homogenized, locally based, perhaps 'hand-made' product.

The desire for reliable quality foods is inherently a product of globalization because it is a reaction to the abysmally low quality standards of mass-produced food, whose manufacture and sale is increasingly controlled by just a few major global corporations.[31] This gives rise to 'a new type of locality, which, in any given area, is the result of the interaction of the various forces that operate from a variety of fields, to confer value upon that space'.[32] Such places are viewed by such agents of economic globalization 'in the entrepreneurial terms of opportunities for investment, infrastructural pluses and minuses, linkages, networks and flows of goods and information'.[33] This, from LEADER's perspective, is in no way contradictory with claiming a clean, green environment as a marketing device.

Frouws (1998), on the other hand, views this type of development strategy as part of what he terms the 'hedonist discourse', which is a view of the countryside originating from the urban elite. He argues that the rhetoric of rural renewal, landscapes of quality and relocalization serve a deeper structural function:

> as a pretext and a licence, under the aegis of contemporary neo-liberalism, for the on-going scale enlargement and industrialisation of agriculture as well as the relentless activities of property developers looking for investment opportunities, which tend to suffocate the very process of renewing the countryside.[34]

While views on this will inevitably differ, among participants and commentators alike, the fact remains that this strategy, and its equivalents elsewhere, will be supported by the EU for many years to come.

Revitalizing the Local Economy II: Local Exchange Trading Systems (LETS)

Contrary to the EU-sponsored, commercial approach of LEADER, one small group of people in west Cork are attempting to revitalize the local economy in an alternative way. This group are involved in the global economic network LETS. This is composed of a network of people who attempt to delink from the international cash-based economy as much as possible. They trade goods and services on a barter basis, among a limited group in the local area. They use their own currency as an alternative to cash, and this is given a local name. Trading accounts are recorded by a central administrator, and then published in the system's directory.[35] They try to maintain wealth within the locality, actively contributing to the economic and social life of their communities. People often choose to offer services to LETS that are not the same as their main paid job, so a teacher might paint houses or a farmer might fix cars.

While LETS is based on an ideology of empowering the local community, it is also very much a product of globalization. Firstly, LETS is primarily the territory of incomers in rural areas like west Cork. Secondly, it is an international phenomenon that has spread as a result of global networking via travel and telecommunications. Without this contact, the seeds of LETS would not have sprouted in so many different locations. The idea of locality has been redefined by LETS, resolutely confident in the strength of the local, yet also closely connected to many other localities on the globe.

LETS generally forms only a small part of a member's overall economic activity, but one that is symbolically significant. It can be seen as a means of establishing control over one's own economic activity in a dehumanized market economy. LETS (Ireland) states that:

> LETS intends to create a community being in touch with one another and gaining self confidence through the discovered demand for individual performance or services by the participants, therefore raising the region's life quality on the whole.[36]

Their website goes into enormous detail regarding the benefits of LETS systems, outlined under the categories of economic, social and environmental benefits. Theirs is a holistic perspective, integrating these three sets of advantages.

LETS systems exist throughout the western world, having originated in Vancouver Island, Canada in 1979. It is significant that LETS was born on an island location during a severe economic depression. Under the leadership of Michael Linton it spread throughout British Colombia in the early eighties.[37] LETS has become popular in Australia, New Zealand, Canada and in every European country. There are 250 LETS in the UK alone, concentrated mainly in the south.[38] There are currently seven LETS operating in Ireland, in west Cork, Cork, Dublin, east Clare, Galway, Mayo and Westport. The first Irish LETS system came into being in Westport, Co. Mayo in 1993.[39] Shortly afterwards, in the mid-nineties, an active LETS system was established in west Cork, on the Beara Peninsula. However, the core of west Cork LETS moved to Bantry, a more central location, in 1997, and the Beara LETS died off completely. It was granted five Fás workers whose job it was to set it up. It is now more properly called Bantry Area Trading System, or BATS, after which the currency is named. One BAT is equivalent to one euro. LETS has been operating in the Bantry area since 1997. The potential local benefits are outlined on their website:

The social benefits of such a system are enormous: interaction between isolated social groups is made possible, and small enterprises can be launched. It is of value to any community in that participation is rewarded, and self-esteem is thus increased.[40]

Despite these ambitious ideals, membership has dwindled. There were over two hundred members at the start, but this number has now depleted and it is relatively stagnant at present, down to a core of forty-one members.

LETS systems are multi-faceted indeed, and it is possible to focus on any one of a number of aspects of their composition. From an economic perspective, LETS has been viewed as a form of 'micro-Keynesianism', where local currency is used to create 'an enhanced multiplier effect'. The effects of LETS extend far beyond economics, becoming 'a form of life politics'. It is an element of a 'DIY' cultural movement that

rather than direct confrontation with the state, engages in cultural innovation, challenging the dominant symbolic codes or frames that give shape and meaning to everyday life.[41]

While LETS certainly appears to form part of something akin to a socio-cultural movement, it is nevertheless somewhat transitory in nature, with members moving in and out of it. One social theorist views the fleeting nature of such groups as an actual resource. In Michel Maffesoli's approach to the study of social movements, the impermanence is seen as a permanent state of affairs.[42] He argues that these days, people are less connected to traditional ties of family, kin and community, and are freer to sign up to active social groupings that he terms 'neo-tribes'.[43]

Issues sometimes emerge that make it necessary for people to come together as neo-tribes to defend their communities. When the threat abates, the groups may also disband, or perhaps some members might regroup in some other ways.

This concept could be applied to many groups, but let us now apply it to LETS in west Cork.[44] Virtually all members thereof are incomers, so it seems to be part of the counter-urban experience that was discussed earlier in the chapter. The Bantry LETS membership list reveals few Irish names. Social class alone seems too crude an instrument to analyse this nebulous group

---

Box 5: Neo-Tribes

Neo-tribes are groups of people whose association is based on emotional connections and shared goals. Most people are no longer hindered by duty or contract in contemporary society, and can express themselves quite freely. These neo-tribes are entirely voluntary in nature, and lead to the formation of temporary communities, whether primarily personal or political. They thrive on what Maffesoli (1996) terms *puissance*, or 'the inherent energy and vital force of the people'. These networks of caring and solidarity aim to make the world a better place, counteracting the alienation felt by many. They provide people with sources of identity, or masks, which make them feel stronger and more included in the broader society. This is in no way negated by the fact that people often flit freely from one to another neo-tribe.

---

of people. Their sense of association is broadly based on ecological and/or socialist principles and a possible sense of isolation and exclusion as 'blow-ins'. They have a substantial amount of each type of capital (economic, cultural and social/symbolic), giving them skills that can be used in their work and political activities. The incomers that are actively committed to LETS are those who have put down significant roots in west Cork. They have bought property, they run businesses, they send their children to school, and generally have no intention of moving. Most of the key members are English, some Dutch or German, and just a few Irish. They are generally articulate and confident, and all of them are socially active in one way or another. They all have been or are currently involved in environmentalist or peace activism. Some are self-employed, some in casual employment, some in community development organizations. Generally, this group of people live according to ecological principles. Most are hard-working and have made brave life-altering decisions to commit to independent counter-cultural lifestyles. LETS is a means of making contact with others of like mind, and a means of counteracting isolation. It creates a kind of manufactured community for this group, which has, perhaps unintentionally, become relatively exclusive to incomers with environmentalist and/or socialist views.

The key members of Bantry LETS are strongly motivated by a politics of economic localization. This may be defined as:

> a plausible way to reverse the instability and insecurity that trade liberalisation has wrought upon the world. The essence of these policies [localisation] is to allow nations, local governments and communities to reclaim control over their local economies; to make them as diverse as possible; and to rebuild stability into community life.[45]

This very sentiment was expressed by one interviewee, who explained his approach thus:

> LETS is only one part of a much bigger picture. Western consumers indirectly contribute to wars. We need to learn to live without polluting and exploiting resources. Mainstream economics is destroying the world. We have to try to localise the economy and reduce food miles.

According to its advocates, then, the economic imperative of LETS involves retaining wealth within the community and not contributing to the massive profits of multinational corporations. This suggests that there is a strong ideological and/or symbolic element to LETS membership, providing a mask of identity that makes people feel comfortable in the context of globalization.

LETS also serves as a basis for network-building among a group of people who are interested in strengthening the local economy and reducing their dependence on the mainstream economy. It has served as a social springboard for involvement in other political groupings. Some members met up years earlier in Earthwatch, for example, and joined LETS afterwards. In recent months, a Cork branch of the Irish Social Forum has been established by some of the Bantry LETS members: Cork Social Forum (CSF).[46] This in turn is part of a global network based on co-operation and solidarity. The issues that have been raised so far at CSF meetings include single father's rights, gay rights, immigrant support, water quality, neutrality and picketing Irish arms factories. This is the type of forum referred to by Melucci, which is independent of government and can create a 'democracy of everyday life'.[47]

Another network that is intimately connected to LETS is the growing number of farmers' markets in Co. Cork and elsewhere throughout the country. People are growing more concerned about the traceability of the food they eat. Since the several food scares experienced in Europe and the US in recent years, the politics of food has become much more sensitive and indeed potentially explosive. As with Fuchsia Brands, producing foods with connotations of the 'local' has now been re-evaluated as a potential developmental niche. Some producers are thus now selling their goods at farmers' markets in order to gain competitive advantage. The environmental benefits of this approach are outlined succinctly in the following, by one of its main Irish advocates, chef Darina Allen:

> Farmers' markets benefit the environment by encouraging sustainable agriculture and small-scale, less intensive production. They reduce the effects of long-distance transport of food and the need for excess packaging.[48]

A very successful national conference on farmers' markets was held in Cork in February 2005, which 250 delegates attended. IFA President John Dillon also showed up to address the meeting, saying that these markets were an important new way of increasing income for a small but growing number of farm families. A new organization has also been established to strengthen the bargaining power of those who sell through the markets, the Irish Food Market Traders Association.[49]

One local cheesemaking company in west Cork is waging its own war at present with the DAF. Sean Ferry and Bill Hogan of West Cork Natural Cheese make the wonderful and widely renowned Gabriel and Desmond cheeses. They make these with raw, unpasteurized milk, which they say is crucial to the flavour and 'integrity' of the cheese. The department wants them, and all other cheesemakers, to pasteurise the milk they use. Hogan claims that they maintain the highest of hygiene standards and no trace of any pathogen has ever been found in their cheese. Obviously, this makes sense as the entire reputation of this small company fundamentally depends upon high standards. They have been engaged in a battle for several years now and have overturned five detention orders of the DAF. The cost of the legal battle is putting them under major strain and they can barely survive, 'hanging by a thread'. Hogan says that 'by promoting and maintaining only corporate food production in Ireland, the state has been missing the point and revealing their bias.[50] With the current relaxation of such hygiene rules at EU level, is it not time for the Irish state to re-assess its priorities?

Having moved to rural Ireland, the incomers involved in such activities have now developed a group identity based on their 'blow-in' status and their common goal of economic localization. Some conflicts with locals are inevitable because their ideals (whether socialist, anarchist, or deep green) often appear to the local population to be Luddite, anti-progress and anti-farmer. This is a key issue that needs to be addressed by this socially active group, before they can gain more widespread social acceptance in the locality. In the long run, however, this appears to be a way for a group who would often have been dismissed as 'hippies' or 'blow-ins' by local people to carve a niche for themselves in the local social structure, and thereby to achieve more social acceptance and integration on their own terms. This is especially the case in the context of a renewed commitment on behalf of the EU to encourage ecological farming practices and overall care for the environment.[51]

It is often only when conflicts emerge that the parties involved are forced to clearly articulate their positions. An example of these 'moments of mobilisation'[52] was seen in recent years in Bantry after an impromptu street market started to become a regular occurrence, especially on the first Friday of every month, which had traditionally been the Fair Day in the town. The local business owners objected to its presence, and a court battle ensued. One of the casual traders represented himself in court and ultimately won the case. As already mentioned, the Fair Day is now a bustling affair and undoubtedly brings more business to the shop, pub and hotel owners who objected to it in the first place. This is a major coup for the 'blow-ins', who have a history of conflict with the locals in the town. Melucci says of such conflicts:

> they can prevent the system from closing in upon itself by obliging the ruling groups to innovate, to permit changes among elites, to admit what was previously excluded from the decision-making arena and to expose the shadowy zones of invisible power and silence which a system and its dominant interests inevitably tend to create.[53]

This counter-urban group, being rich in cultural capital, are likely candidates for such forms of cultural innovation. This group take their personal role in community building extremely seriously. They blend modern and traditional goals to create what Maffesoli terms a 'dynamic rootedness'.[54]

Conclusion

West Cork is a meeting-place between different sets of views and interests that are constantly, either implicitly or explicitly, in competition with each other. Overall, this is a region with a relatively poor economic base. Farming is declining in importance, and employment is to be found these days in the service sector, retailing, tourism and construction. The area's physical beauty and coastal location will ensure its continuing popularity with tourists, provided that the worst excesses of property development are avoided. Local actors operate in interpreted cultural worlds that are shaped by the interaction between global and local levels. The present and future of regions such as west Cork are shaped primarily by EU and government policy for peripheral regions, and the local people who live there do not have much choice but to submit to this policy, inserting themselves in the broader picture in whatever ways they can.

As we have seen, some have been very innovative in their attempts to literally make a place for themselves. The Fuchsia Brands initiative has constructed the idea of an Irish gourmet region. Only a retrospective view will tell the real impact of such LEADER projects on the communities in which they operate. While specific research needs to be conducted in this area, one would expect that it is those who are endowed with economic, cultural and symbolic capital that dominate this type of initiative in west Cork. They are probably numbered among the group that Eipper, in his seminal study of power relations in Bantry, *The Ruling Trinity* (1986), termed 'the influentials'. These are the established bourgeoisie who are endowed with social status and power. Similar criticisms have been levelled at other LEADER initiatives by other commentators. It has been argued, firstly, that it is usually the better-off who benefit most from LEADER funding and the majority of rural-dwellers remain unaffected; secondly, that 'empowering' local communities to provide services for themselves is a means of cost-cutting on behalf of the welfare state, with local volunteers thus becoming a source of cheap or free labour; and thirdly, that it tends to professionalize local development and incorporate local initiatives into a supra-national bureaucratic framework.[55] Overall, there seems to be a common suspicion that LEADER may tend to promise more than it actually delivers in terms of consolidating local development strategies and creating employment for those who need it most.

While LETS is not exactly thriving at the moment in west Cork, it nevertheless has spread its wings and given rise to other types of contact between people where different issues are addressed. This group of people are socio-cultural innovators whose actions and strategies do not always appear coherent to established social actors. While appearing to be transient and disorganized, they nevertheless are a strong indicator of the need for new intellectual spaces to be created in rural Ireland.

LETS forms part of a growing constellation of groups who are concerned with the future of the Irish countryside. Its members are a neo-tribe who connect with each other on the basis of shared ecological values. This places them among a coterie of other similar groups who previously may have been perceived as 'left-field' and countercultural, but who now have the ear of bureaucrats and politicians at the highest levels. They use LETS as a social springboard from which to launch other economic, social and political activities. Their enthusiasm is infectious as they signify for rural areas somewhat of a renaissance. However, Maffesoli warns that the activities of neo-tribes are not always the most effective:

> This is a thoroughly tragic perspective, which is aimed less at changing the world than getting used to it and tinkering with it.[56]

This 'tinkering', however, can produce fairly profound local effects by opening peoples' minds and instilling some hope where despondency may previously have been the order of the day. Lifestyles that in the past were profoundly counter-cultural and *avant-garde* are now becoming part of the mainstream, as mere hobbies become thriving businesses. When small business owners and local development aficionados engage with each other a lot is learned in the process, empowering both parties with new talents and skills. In this way, the goals of LETS and other small businesses operating in the area may end up dovetailing rather than conflicting. The outcomes of such encounters may well lead to a much more sustainable form of economic development than that hitherto offered by multinational corporations.

Conclusion

Pat Murphy of Ballina, Co. Mayo has recently embarked upon an interesting venture: a 'mobile farm' that brings a selection of farm animals around to visit primary schools in the region. Schoolchildren can have hands-on contact with the animals and even try their hand at milking.[1] The fact that this is a novelty is an indicator of just how urbanized Irish society has become. The public, and especially children, now see food as a neatly wrapped package on a supermarket shelf rather than having much connection with rural life. Considering the growing epidemic of child obesity, this is especially worrying. Is this type of development a portent of the future? Is the study of the social composition of the Irish farming population to become the territory of historians rather than sociologists?

Farmers are generally extremely pessimistic about their future. They are very much aware of the extent of their EU dependence and fearful of prices dropping for the commodities they produce, and of the termination of the quota system. Their fears are certainly not unfounded. Nearly all of the farmers interviewed mentioned this structural problem, some typical quotes being:

> If the EU stays pumping money towards us, we'll survive. If not, we'll die. It's as clearcut as that really. If quotas go and if milk

drops to 60 or 70p, with no subsidy of any type, you'd be talking about automatic disqualification.

In twenty years' time, you're going to have a lot of other countries supplying food products into Europe, like India and Eastern Europe, so I see a bad future. Unless tourism, or something else will bring up the security of farming.

This can create a certain amount of despondency among farmers who lack the confidence to try new things. Many would prefer to get out altogether rather than to take bigger risks. This worries John Feehan:

A certain *idealism has been lost*, and this loss is often linked to a quiet despair which grows into a cancer below the surface, especially among the rural young.[2]

Survival of the fittest is now the dominant philosophy within Irish farming. At a global level, the choices faced by small farmers are limited indeed and they 'either bootstrap their way into a market niche through some non-farm enterprise, or provide a new low(ered)-cost labor force for a global firm'.[3] Within the previously protected arena of the EU, this is now also the case. Franz Fischler, the EU Commissioner for Agriculture, recently asserted that

the outlook for many agricultural commodities is not very promising in Europe ... [we need to] deepen and extend the 1992 CAP reforms, [involving] ... a stepwise reduction of 30 per cent in intervention prices.[4]

This policy change means that price support mechanisms upon which farmers are currently relying for income can certainly not be guaranteed in the long term. The production of the usual commodities is no longer adequate for survival in this period of 'structural adjustment'. Instead, he says that the countryside is to serve 'multi-functional roles'. Some of the directions Fischler suggests are 'the production of renewable raw material for non-food purposes or the energy sector, rural tourism, marketing of high-quality produce or the preservation of our cultural heritage'. While this might sound exciting to those with the resources to innovate, it could sound quite threatening to small farmers who have been farming for generations and who have few other skills besides the intimate knowledge of their farms.

Diversification stretches available financial resources and a policy of retrenchment and cost-saving probably appears to be more sensible than expensive investment in new ventures. Fear was expressed by the farmers about this new multi-functionality for the land:

> I think that the environmentalists are going to start winning in all this because the farming lobby is getting smaller. The country is going to become a place for holiday homers and farmers have very little control over it.

> A lot of people are leaving the countryside and it's being used as a kind of a playground, just to come down at weekends and in the summer.

While diversification is undoubtedly required, there is an inadequate system of training in place to get farmers beyond producing the traditional commodities.[5] People have tried many different initiatives, with varying degrees of success. As discussed in chapter seven, alternative food production is burgeoning in some areas. In others, tourist ventures of various kinds have been tried, and one of the latest activities to be encouraged is floristry. There is a market for cut flowers and training schemes are in place for those interested. It is emphasized by the DAF that the enlargement of the EU (as of May 2004) has expanded the market for Irish goods and that farmers should be optimistic in this regard. The official line is that the accession of the 10 new countries presents more opportunities than threats to Irish farmers, except in a few sectors, like mushrooms, potatoes, pigmeat and cereals. Farmers therefore need to be supported by state agencies in order to diversify and to satisfy these new markets. This may be diverted through community initiatives, many of which are already doing sterling work.

This book has been a sociological investigation of social change in rural Ireland since accession to the EEC in 1973. The modernization of Irish farming created a substantial social gap between the smaller farmer and the larger farmer. The former were dismissed by the agents of modernization as traditional and marginal, while the latter benefitted greatly from the EEC/EC/EU pricing mechanism since the seventies. The CAP embodied a productivist mission within European agriculture, which appeared to be challenged by the adoption of agri-environmental measures. This restructuring began in the eighties and continued into the nineties, when EU policy-makers were anxious to reduce the amount of production-related payments made to farmers. They attempted to redesign farming policy by marrying the concerns about financing over-production of agricultural produce, rural poverty and depopulation, and environmental conservation. At this time, it was politic to frame interventions addressing the first two of these issues in terms of the third. It gave the impression in the public eye that these problems were being actively addressed. The proportion of the overall EU budget spent on agriculture is now at 46.5 per cent, reduced from *c*.70 per cent in the late eighties, but it is still the largest drain on the budget.

REPS is the single most important non-price initiative to emerge in Ireland in recent years, with one third of Irish farmers having joined its ranks. The dominant actors in agriculture, those in possession of large amounts of the different types of capital, have wholeheartedly embraced REPS because it does not prevent large farmers from becoming even larger and intensifying their production further. It is a means of distributing a form of welfare to smaller farmers, which supplements their low income levels, and simultaneously reduces them to managers of a manicured landscape for the consumption of urban dwellers.

Meanwhile, in the productive farming zone of the east and south-east of the country, large intensive farmers are still primarily motivated by the large sums that are to be earned from EU prices, especially for milk. Wealth is still primarily accumulated in farming through the EU pricing mechanism and productivist agriculture. The majority of smaller farmers have now been allocated the role of producing environmental goods in a multi-functional rural arena. It has the effect of zoning land and people with low levels of productivity out of producing agricultural goods and into producing environmental goods. This could mean that, simultaneously, agricultural production can become even more concentrated in fewer hands within the farming élite and the green lobby can point to their own successes in setting aside the rest from intensive production. REPS constitutes at best, in the words of several interviewees, 'a handy bonus' for drystock, sheep and small dairy farmers. Meanwhile, in environmental terms, those who produce the most still pollute the most. Those who have joined REPS are generally extensive farmers and are not responsible for large amounts of slurry runoff or nitrates in streams and rivers anyway, except perhaps in exceptional cases of carelessness. At best, it encourages smaller farmers to, as many of them said, 'tidy up the place' or 'clean up their act'. It is also a vehicle for the social control of the smaller farmer by the state.

The bureaucratization that is endemic to REPS is an invaluable tool for maintaining an ideological distance between the power-holders and their subjects. The complexity of the specifications sets up an unequal power structure where the farmers are heavily dependent on the officials to interpret them. It is a top-down policy in which farmers are entreated to change their practices according to a specific expert scientific worldview. It is presented to farmers as a *fait accompli*, with no opportunity for them to make an input into its design. The scheme would be improved by the integration of farmers' views into its design. If more democratic forum existed where their concerns could be aired, it could only be of benefit to the long-term future of environmentally sound agriculture. Farming people need to be consulted in order to hear the many constructive suggestions that they are capable of making. REPS constitutes more of a cosmetic initiative than a radical one, a type of reform that is more mechanical than moral, and an approach to the agri-environment that is more technical than ethical.

The countryside is also subject to other types of influences, meaning that an increasing amount of land is zoned on the one hand for housing development and on the other, for ecologically preserved areas. The conflicts that occur around these issues need to be handled by implementing a sensible, balanced regional development policy without fear or favour for any one group or individual over another. It is also emphasized that rural and urban development policies need to work in tandem, because there is a direct relationship between the problems caused by depopulation in the countryside and the congestion, pollution and housing problems in the major cities.

In terms of sociological analysis, a constructionist approach was advocated in this book. The employment of Bourdieu's theory of practice ordained that one could not analyse social change in rural Ireland without acknowledging the various contributions of the different social actors involved. It is argued that these contributions need to be analysed alongside the underlying mechanisms of capitalism. It is unsatisfactory to view small farmers, for example, as passive in the face of the market. Instead, these forces are 'processed' by small farmers and peasants themselves, 'integrat[ing] them into their own farming strategies, and in this sense retain[ing] a degree of independent decision-making', thereby retaining a 'relative autonomy'[6] in the face of global capitalism.

Social reality is viewed in this research as an unequal power struggle between sets of actors who are proponents of different sets of claims, or story-lines. Each set of actors possess different types and amounts of social resources, or capitals, which govern the amount of influence they can exert over others. Each group, and individual within a group, has the social structure inscribed in their habitus. It is the operation of this habitus that makes the achievement of radical social change so difficult, because the dominant tendency is towards reproduction rather than rupture. This approach combines analysis of socio-economic and socio-political power structures with a detailed look at how the subjects themselves *read* the situation, ascertaining whether or not they actually succumb to that structural power. We can never presume a linear path between the original design of any social policy and its ultimate implementation by real women and men in concrete social settings.

Notes

INTRODUCTION

1 Throughout this book, I will term this region the EU broadly with reference to developments that occurred during and since the nineties, the EC in the eighties and the EEC in the seventies.
2 For the full quotation, see Marx, 1984: 360.
3 Mills, 1959: 14.
4 See Harvey and MacDonald, 1993.
5 Hajer, 1996: 247.
6 Bird, 1987: 258.
7 Walsh, 1998: 14–15.
8 The key work here is Bourdieu's *Outline of a Theory of Practice* (1977).
9 Bourdieu, 1987: 2.
10 Calhoun, 1993: 72.
11 For a good collection of excerpts from many of the key readings on the subject, see Lechner and Boli, 2000.
12 Giddens, 1990: 64.
13 Fröbel *et al*, 1980.
14 Jameson, 1984.
15 Offe, 1984.
16 Harvey, 1989.
17 Robertson, 1992.
18 See Crowley, 1997.
19 Allen, 2000: 186.
20 See www.nologo.org
21 See Chomsky, 2000.
22 See D. Della Porta, H. Kriesi and D. Rucht, 1999; R. Cohen and M. Rai, 2000.
23 Kalb, 2000: 7.
24 Massey, 1994: 277.
25 Taken from P. Muldoon (ed) *The Faber Book of Contemporary Irish Poetry* (London 1986).
26 Quinn, 2003: 304.
27 Heaney, 1990: 154.
28 McMichael, 1996: 50.
29 Gribaudi, 1996: 73.
30 Massey, 1994: 154.
31 Whelan, 1993: 10.
32 Orme, 1970.
33 Whatmore, 1990: 253.
34 When approaching farm families about an interview, I asked to

speak to the person who knew
most about REPS. This was
sometimes the man, sometimes
the woman, and indeed sometimes both.

1. THE CAP AND SOCIAL INEQUALITY

[1] Newby, 1978: 15.
[2] Commins, 1990; Ward, 1993.
[3] Buttel, 1994: 20.
[4] Commins, 1990: 45.
[5] Clark and Lowe, 1992: 17.
[6] Tovey, 1997: 132.
[7] Euro.Comm., 1980: 1.
[8] Bonanno, 1991: 552.
[9] Commins, 1995: 181.
[10] Bonanno, 1993: 349; Palmer, 1988: 93.
[11] Palmer, 1988: 94.
[12] Bonanno, 1990: 163.
[13] Euro.Comm., 1996: 3.
[14] Breen *et al.*, 1990: 198.
[15] Dooney, 1988: 5.
[16] Matthews, 2000: 85.
[17] See Arce, 1993.
[18] Tovey, 1997: 131.
[19] Leeuwis, 1989.
[20] *Cork Examiner*, 26 Oct. 1995.
[21] Breen *et al.*, 1990: 194.
[22] Scott, 1995: 107.
[23] Bonanno, 1990: 177.
[24] Coughlan, 1980: 127.
[25] Cox, 1985: 12.
[26] Breen *et al*, 1990: 196.
[27] Commins, 1995: 186.
[28] Cox, 1985: 23.
[29] Breen *et al.*, 1990: 199.
[30] *Irish Times* editorial, 17 Nov. 1998.
[31] Matthews, 2000: 80.
[32] See Cawley, 1999.
[33] Scott, 1995: 105.
[34] *Irish Times*, 29 May 1997.
[35] O'Hara, 1986: 45.
[36] Commins, 1995: 186.
[37] Nolan, 1998.
[38] When the pound sign (£) is used it refers in all cases to Irish pounds, which was the currency in Ireland until the introduction of the euro in January 2002.
[39] Connolly, 2002.
[40] Frawley and Commins, 1996; Connolly, 2002.
[41] Goodman and Redclift, 1981: 7.
[42] Shucksmith, 2003.
[43] EUROSTAT, 1997.
[44] Commins, 1995: 197.
[45] Bonanno, 1993: 558.
[46] Symes, 1992: 195.
[47] Eipper, 1986: 22.
[48] O'Keeffe, 2004.
[49] Donnelly and Crosse, 2000.
[50] Feehan, 2003: 522.
[51] Connolly, L., A. Kinsella, and G. Quinlan, 2003.
[52] Matthews, 2000: 58.

2. SURVIVAL STRATEGIES IN HARD TIMES

[1] Goodman and Redclift, 1981: 3.
[2] Marx, 1984: 378.
[3] Goodman and Redclift, 1981: 5.
[4] Kerblay, 1984: 151.
[5] Chayanov, 1925: 6.
[6] van der Ploeg, 1994: 17.
[7] Bourdieu, 1987: 2.
[8] Calhoun, 1993: 72.
[9] Bourdieu, 1977: 72.
[10] Bourdieu, 1984: 471.
[11] Wacquant, 1998: 220.
[12] DiMaggio, 1979: 1461.
[13] Bourdieu, 1977: 166.
[14] *Ibid.*: 79.
[15] *Ibid.*: 82.

[16] Wacquant, 1992: 23.
[17] Bourdieu, 1987: 4.
[18] FMG, 1990: 207–8.
[19] Bourdieu, 1977: 188.
[20] Bourdieu, 1985: 731.
[21] Bourdieu, 1977: 179–80.
[22] Bourdieu, 1991: 23.
[23] Ortiz, 1984: 332.
[24] CSO, 1997.
[25] Commins and Keane, 1994: 39.
[26] Hannan and Commins, 1992.
[27] Donnelly, 2003.
[28] Goodman and Redclift, 1981: 19.
[29] van der Ploeg, 1994.
[30] Commins and Keane, 1994: 77.
[31] Kinsella *et al.*, 2000: 485.
[32] Quoted in Young, 2003.
[33] Newby, 1978: 19.
[34] Ortiz, 1984: 333.
[35] Arensberg and Kimball, 2001.
[36] Storey, 1994.
[37] Tovey *et al.*, 1996: 10.
[38] Jackson and Haase, 1996: 82.
[39] Hannan and Commins, 1992.
[40] MacLaughlin, 1997(b): 136.
[41] Hannan, 1979: 201.
[42] MacLaughlin, 1997(a).
[43] Ó Gráda, 1973: 56.
[44] The best source on this issue is O'Hara, 1998.
[45] Coulter, 1993: 33.
[46] Wilson, 1985; 1989.
[47] Dooney, 1988: 51.
[48] Eipper, 1986; Wilson, 1989.
[49] *Farmer's Journal*, 3 Feb. 2004.
[50] Chubb, 1991: 57.
[51] Palmer, 1988: 92.
[52] Thompson, 1968: 10.
[53] Long, 1992: 20.

3. CHANGING THE GOAL-POSTS

[1] Euro. Comm., 1980.
[2] Deverre, 1995: 232.
[3] Storey, 1994.
[4] Commins, 1990: 66.
[5] Clark and Lowe, 1992.
[6] Euro.Comm., 1987: 9.
[7] See Commins, 1990; Ward, 1993.
[8] Whitby, 1996: 233.
[9] Vail *et al.*, 1994: 34.
[10] Euro.Comm., 1985: 20.
[11] Brunt, 1990: 22.
[12] Euro. Comm., 1985: 21.
[13] Scott, 1995: 113.
[14] Commins, 1990: 66.
[15] Vail *et al.*, 1994: 241.
[16] Ward, 1993: 357.
[17] Bonanno, 1990: 174.
[18] General Agreement on Tariffs and Trade, the precursor of the World Trade Organization (WTO).
[19] Philips, 1990: 167.
[20] Bonanno, 1990: 174.
[21] Whatmore, 1990: 252.
[22] Euro. Comm., 1987: 9.
[23] Commins, 1990: 65.
[24] Vail *et al.*, 1994: 35.
[25] Commins, 1995: 189.
[26] Buttel, 1994: 29.
[27] Commins, 1990: 71.
[28] See Marsden *et al.*'s benchmark 1993 publication *Constructing the Countryside*.
[29] Marsh, 1991: 50.
[30] Commins, 1990: 71.
[31] Symes, 1992.
[32] Fischler, 1997: 36.
[33] Ward, 1993: 362.
[34] This convenient formulation comes from Marsh, 1991.
[35] Potter, 1998: 98.

[36] Euro. Comm., 1996: 4.
[37] Lowe, 1992: 7.
[38] Deverre, 1995: 238.
[39] Baldock and Lowe, 1996: 23.
[40] Commins, 1990: 66.
[41] Ward, 1993: 357.
[42] Whitby, 1996: 227; Baldock and Lowe, 1996: 17.
[43] Euro. Comm., 1992: 86–7.
[44] Green Group, 1993: 91.
[45] Nugent, 2000: 26.
[46] Cypher and Dietz, 1997: 343.
[47] This development would sadden one artist who moved to live in west Cork. Of the various improvized stop-gaps one typically saw in the region, he writes: 'All these small manifestations of care interest me – the well-placed Commer van wedged between gateposts, the carefully angled bed-end lashed to a sapling with binder twine – and seem worthwhile subjects for drawings' (Lalor, 1990: 5). This shows that what may serve as an indicator of backwardness and carelessness to the agricultural scientist, may appear as a worthy subject of creative attention to the artist.
[48] Shirley, 1998.
[49] This review was conducted in 2002. See www.teagasc.ie
[50] Evans and Coen, 2001: 16.
[51] See www.teagasc.ie
[52] Caslin, 2003.
[53] Dineen, 2003.
[54] Matthews, 2000: 82–3.
[55] Editorial, *Farmer's Journal,* 24 Jan. 2004
[56] MacConnell, 2004(a).

4. GREEN CAPITALISM

[1] Connelly and Smith, 1999: 58.
[2] Harvey, 1996: 378–80.
[3] WCED, 1987: 43.
[4] Redclift, 1987.
[5] Redclift, 1992: 25.
[6] Redclift, 1987: 35.
[7] Escobar, 1996: 49.
[8] Euro Comm., 1996: 13.
[9] Euro.Comm., 1997: 103.
[10] *Ibid.*: 111.
[11] *Ibid.*: 127.
[12] Frouws, 1998: 60.
[13] Eder, 1996: 217.
[14] DAF, 1997: 9.
[15] Hannigan, 1995: 20.
[16] Mol, 1996: 315.
[17] Arce, 1993: 2.
[18] Bourdieu, 1977: 21.
[19] Buttel and Taylor, 1992: 214.
[20] Hajer, 1996: 246.
[21] Portela, 1994: 39.
[22] Goodman and Watts, 1994: 30.
[23] Symes, 1992: 206.
[24] McEvoy, 1999: *x.*
[25] Harvey, 1996: 374.
[26] McEvoy, 1999: *x.*
[27] Scott, 1995: 120.
[28] Symes, 1992: 195.
[29] Portela, 1994: 47.
[30] Arce, 1993: 167.
[31] Mooney, 1997: 8.
[32] Shutes, 1991: 14.
[33] Symes, 1992: 203.
[34] REPS III now pays out a 10 per cent incentive to this group, however.
[35] Midgeley, 1997: 104–6.
[36] Offe, 1984: 156.
[37] Finn, 2003: 21.
[38] *Ibid.*
[39] Potter, 1998: 103.
[40] O'Sullivan, 1999.
[41] Pocock, 2004.

[42] Euro.Comm., 1989: 1.
[43] Lee, 1999: 15.
[44] Marsh, 1991.
[45] Murphy and Lally, 1998: 91.
[46] EPA Pamphlet, 1999.
[47] Evans and Coen, 2001: 15.
[48] Murphy and Lally, 1998: 75.
[49] Donnelly, 2003.
[50] Keohane, 1998: 91.
[51] ENFO, 2000.
[52] See www.europa.eu.int/eur-lex for full text of EU Directives.
[53] One scientist, Dr J. Mulqueen of NUI Galway, argues that nitrates do not cause stomach cancer, but are actually good for us (*Farmers Journal,* 21 Feb. 2004, p. 11).
[54] Culleton and Dillon, 2004.
[55] Farmer's Journal, 5 July 2004.
[56] Pocock, 2004.
[57] Pocock, 2004(a).
[58] Mooney, 2004.
[59] Editorial, *Farmer's Journal,* vol.57 (6) 7 Feb. 2004.

5. LANDSCAPE AND HERITAGE

[1] Bell, 1993: 1.
[2] Hannigan, 1995: 39.
[3] *Ibid.:* 185.
[4] Hajer, 1996: 257.
[5] Bird, 1987: 255.
[6] For detailed sociological discussion of this argument, see Burningham and Cooper, 1999.
[7] See Hajer, 1996: 258.
[8] Evans, 1957: 18.
[9] His influence extended into geography, history, anthropology and folklore, and the numerous interstices between these disciplines.

[10] Bell, 1993: 20.
[11] Daniels and Cosgrove, 1988: 8.
[12] Feehan, 2003: 510.
[13] Department of the Environment, Heritage and Local Government, *Living With Nature* Dublin 2003.
[14] See www.europa.eu.int/eur-lex for full text of EU Directives.
[15] See www.birdwatchireland.ie for more information
[16] McDonald, 2004(c).
[17] Reid, 1998.
[18] This is in Minister Cullen's own constituency.
[19] Staunton, 2004.
[20] McDonald, 2004(c).
[21] Viney, 2003: 14.
[22] *Ibid.*
[23] *Ibid.:* 15.
[24] Feehan, 2003: 510.
[25] Aalen, 1997: 257.
[26] Feehan, 2003: 511.
[27] Feehan, 1997: 583.
[28] Tovey,1994: 214.
[29] *Ibid.:* 212.
[30] Feehan, 1997: 587.
[31] Mooney, 1998.
[32] *Farmer's Journal,* 'A SAC Full of Trouble?' Vol.52 (27) 1 July 2000.
[33] Maguire, 1997: 3.
[34] Viney, 2003: 12.
[35] See Urry, 1990.
[36] See McGrath, 1996.
[37] Cabot, 1999: 428.
[38] See Tovey, 1996.
[39] One exception is the work of Byrne *et al.,* 1993.
[40] Ó Neachtain, 2004.
[41] Letter to *Farmer's Journal,* 7 Feb 2004.
[42] Herman, 2004.
[43] McNally, 2004.
[44] See www.keepirelandopen.org

45 Garland, 1998.
46 MacConnell, 2004.
47 Lavery, 2004.
48 Cooney *et al.*, 2000: 22–3. The collection in which this is found is the single best source on Irish heritage issues: Neil Buttimer *et al.*'s *The Heritage of Ireland.*
49 *Ibid.*: 19.
50 McDonald, 2005.
51 See www.heritagecouncil.ie
52 Roche, Dick 'Downright distortion drives Tara motorway controversy' *Irish Times* 21 May 2005.
53 Gartland, F., 'Majority want M3 routed outside Tara, survey finds' *Irish Times* 20 Aug 2005.
54 See www.heritagecouncil.ie/news/M3_hill_of_tara.html
55 McDonald, 2005.
56 O'Toole, Fintan 'M3 and the uglification of Ireland' *Irish Times* 17 May 2005.
57 Mr Salafia has a blog (hilloftara.blogspot.com) with updates on developments.
58 Feehan, 2003: 526.

6. CONTRASTING RURAL VISIONS

1 McCarthy, 2000: 117.
2 Cullen, M., 'Recognising the Needs of Rural Ireland' *Farmers Journal* 28 June 2004.
3 Devlin, 2004.
4 Department of the Environment and Local Government, *The National Spatial Strategy 2002–2020: People, Places and Potential* (Department of the Environment and Local Government, Dublin 2002).
5 Haynes, 2004: 31.
6 McDonald, 2001.
7 McDonald, 2004.
8 Cabot, 1999: 435.
9 McDonald, 2004(a).
10 *Ibid.*
11 Smith, 2004.
12 McDonald, 2004(a).
13 *Ibid.*
14 O'Dwyer, 2004. See also www.whoownsscotland.org.uk and www.caledonia.org.uk/land for discussion of land ownership and land reform in Scotland.
15 McDonald, 2004(b).
16 Mullane, 1998: 47.
17 See www.antaisce.org
18 RRI abandoned its website because it was attracting too many responses from all over the world, which were outside its remit.
19 RRI brochure *Moving to the Country ... For the Sake of the Children.*
20 Jim Connolly, interview with the author, Sept. 2004.
21 McDonald, 2004(a).
22 See Monbiot, 2000, chapter 5.
23 Monbiot, 2004.
24 To read the perspective of Via Campesina, a powerful international farmers movement that challenges these trends, see www.viacampesina.org
25 There were also payments for those with under 3 has and more than 1 ha under fruit and vegetables (which would be quite rare in Ireland). Most organic fruit and vegetables are imported, so this is understandably an attempt to encourage more production in this area. There was never a strong tradi-

tion even of conventional production of fruit and vegetables
in Ireland. Horticultural products account for only 3 per cent
of overall Irish agri-food and
drink exports, whereas dairy
products account for 24 per
cent and beef and live animals
21 per cent.

26 Friemann, 2004.
27 Crowley, 2000.
28 www.theorganiccentre.ie
29 Tovey, 1999: 39.
30 Friemann, 2004.
31 Phipps, 1999.
32 Friemann, 2004.
33 Tovey, 1999: 55.
34 *Ibid.*: 57.
35 This data comes from a recent
Bord Bia report on Irish organics quoted in Friemann, 2004.
Having accessed the Bord Bia
website (www.bordbia.ie), I
found that this report was not
accessible to the public and
costs €500 to buy, so I could
not see the original.
36 A list of venues and times of
farmers' markets all over the
country can be found on
www.bordbia.ie
37 See Sage, 2003.
38 See www.slowfood.com
For the Irish scene, see
www.slowfoodireland.com
39 Friemann, 2004.
40 Marsh, 1991.
41 See www.ballybrado.com
42 Schulte *et al.*, 1998.
43 *Ibid.*
44 *Farmers' Journal*, 26 July 2003.
45 Cadogan, 2003.

7. WEST CORK

1 See Eipper, 1989; Allen and
Jones, 1990.
2 McCarthy, 2000: 9.
3 MacLaughlin, 1994: 434.
4 Commins and Frawley, 1994.
5 Massey, 1994: 155.
6 Whelan, 1993: 11.
7 Donnelly, 1975.
8 Dickson, 1993.
9 Interview, Schull, 1999.
10 *Prime Time*, RTÉ, 24 Feb. 1998.
11 Sheridan, 2004.
12 See Cloke, 1985; Halfacree,
1997.
13 Eipper, 1986: 134.
14 Lalor, 1990: 22.
15 Forsythe, 1980: 287.
16 Phillips, 1986.
17 Peace, 1986: 114.
18 Martin, 1999: 175.
19 West Cork LEADER Co-Op,
1996.
20 O'Rielly, 2003: 172.
21 See www.fuchsiabrands.com
22 Interview, Dec. 2002.
23 Slater, 1993: 10.
24 Commins and Frawley, 1994: 37.
25 See www. westcorkleader.ie
26 See Breathnach, 1992.
27 O'Rielly, 2003: 186.
28 Sage, 2003: 54.
29 Ilbury *et al.*, 2001: 30.
30 Strassoldo, 1992: 40.
31 For further discussion of these
issues, see Bonanno *et al.*, 1994;
McMichael, 2000; Hendrickson
and Heffernan, 2002.
32 Mormont, 1990: 32.
33 MacLaughlin, 1997: 193.
34 Frouws, 1998: 64.
35 Other authors have provided
enormous detail regarding the
minutiae of the economic

workings of LETS that do not
need to be reproduced here. See
Douthwaite, 1996; Lee, 1996;
Offe and Heinze, 1992; Thorne,
1996; Williams, 1996.

[36] Learn more about LETS Ireland
at www.poptel.org.uk/aries/envi-
ronet/archive/msg00078.html

[37] Offe and Heinze, 1992: 86.

[38] See www. LETS-Linkup.com for
enormous detail about LETS
internationally, including contact numbers and addresses for
each one.

[39] Douthwaite, 1996: 77.

[40] www.westcorkweb.ie/lets/

[41] Purdue *et al.*,1997: 647.

[42] This argument is explained fully
in his book *The Time of the
Tribes*, 1996.

[43] For those who wish to study the
archaeology of this key idea, see
Hetherington (1992), who
explains that it may be traced to
a strong influence by Schmalenbach's groundbreaking essay
'Communion – A Sociological
Category' first published in
1922 (Schmalenbach 1977).
This was a critique of Toennie's
classic distinction between
Gemeinschaft (community) and
Gesellschaft (society). Schmalenbach added the category of
'communion', which refers to
those forms of emotional association in which people voluntar-
ily engage.

[44] For a longer discussion of
Bantry LETS, see Crowley, 2004.

[45] Hines, 2000: 29.

[46] See www.irishsocialforum.org
for more information.

[47] Melucci, 1988: 259.

[48] Allen, 2003: 6.

[49] See www.irelandmarkets.com

[50] Hogan, Bill 'The Irish Cheese
Wars' *Seilide: Slow Food Ireland
Magazine,* 7 June 2005.

[51] See Crowley, 2003.

[52] Melucci, 1988: 248.

[53] *Ibid.*: 254.

[54] Maffesoli, 1996: 33.

[55] See Martin, 1999 and Storey,
1999.

[56] Maffesoli, 1996: 51.

CONCLUSION

[1] Shiel, 2003.

[2] Feehan, 2003: 524.

[3] McMichael, 1996: 47.

[4] Fischler, 1997: 35.

[5] One exception is www.farmop-
tions.ie

[6] Long, 1986: 19.

Bibliography

Aalen, F.H.A., 'Management of the Landscape' in Aalen, F.H.A., K. Whelan and M. Stout (eds), *Atlas of the Irish Rural Landscape* (Cork 1997).

Allen, D., 'Farmers' Markets – Every Town Should Have One', *Farmer's Journal*, 56, 48 (2003).

Allen, K., *The Celtic Tiger: The Myth of Social Partnership in Ireland* (Manchester 2000).

Allen, R., and T. Jones, *Guests of the Nation* (London 1990).

Appadurai, A., 'Disjuncture and Difference in the Global Cultural Economy' in F.J. Lechner and J. Boli, *The Globalization Reader* (Massachusetts 2000).

Arce, A., *Negotiating Agricultural Development: Entanglements of Bureaucrats and Rural Producers in Western Mexico* (Wageningen 1993).

Arensberg, C., and S.T. Kimball, *Family and Community in Ireland*, 3rd edn (Clare 2001).

Baldock, D., and P. Lowe, 'The Development of European Agri-Environment Policy' in M. Whitby (ed.), *The European Environment and CAP Reform: Policies and Prospects for Conservation* (Wallingford 1996).

Bell, D., 'Framing Nature: First Steps into the Wilderness for a Sociology of the Landscape', *Irish Journal of Sociology*, 3 (1993), 1–22.

Billaud, J-P, K. Bruckmeier, T. Patricio and F. Pinton, 'Social Construction of the Rural Environment: Europe and Discourses in France, Germany and Portugal' in de H. de Haan, B. Kasimis and M. Redclift (eds), *Sustainable Rural Development* (Aldershot 1997).

Bilton, T. *et al.*, *Introductory Sociology* (Basingstoke 2002).

Bird, E.A.R., 'The Social Construction of Nature: Theoretical Approaches to the History of Environmental Problems', *Environmental Review*, 11, 4 (1987), 255–64.

Bonanno, A., 'Social and Economic Consequences of the EC Agricultural Policy' in A. Bonanno (ed.), *Agrarian Policies and Agricultural Systems* (Boulder 1990).

Bonanno, A., 'From an Agrarian to an Environmental, Food, and Natural Resource Base for Agricultural Policy: Some Reflections on the Case of the EC', *Rural Sociology*, 56, 4 (1991), 549–64.

Bonanno, A., 'The Agro-Food Sector and the Transnational State: The Case of the EC', *Political Geography*, vol 12, 4 (1993), 341–60.

Bonanno, A. *et al.* (eds), *From Columbus to Conagra: the Globalization of Agriculture and Food* (Kansas 1994).

Bourdieu, P., *Outline of a Theory of Practice* (Cambridge 1977).

Bourdieu, P., *Distinction: A Social Critique of the Judgement of Taste* (London 1984).

Bourdieu, P., 'The Social Space and the Genesis of Groups', *Theory and Society*, 14 (1985), 723–44.

Bourdieu, P., 'What Makes a Social Class? On the Theoretical and Practical Existence of Groups', *Berkeley Journal of Sociology*, 32 (1987), 1–17.

Bourdieu, P., *Language and Symbolic Power* (Cambridge 1991).

Breathnach, P., 'Employment in Irish Tourism: A Gender Analysis', *Labour Market Review*, 3, 2 (1992), 15–26.

Breen, R. *et al.*, *Understanding Contemporary Ireland: State, Class, and Development in the Republic of Ireland* (Dublin 1990).

Brunt, B., *Western Europe: A Social and Economic Geography* (Dublin 1990).

Burningham, K., and G. Cooper, 'Being Constructive: Social Constructionism and the Environment', *Sociology*, 33, 2 (1999), 297–316.

Buttel, F.H., 'Agricultural Change, Rural Society and the State in the Late Twentieth Century – Some Theoretical Observations' in D. Symes and A.J. Jansen (eds), *Agricultural Restructuring and Rural Change in Europe* (Wageningen 1994).

Buttel, F.H., and P. Taylor, 'Environmental Sociology and Global Environmental Change: A Critical Assessment', *Society and Natural Resources*, 5 (1992), 211–30.

Byrne, A., R. Edmondson and K. Fahy, 'Rural Tourism and Cultural Identity in the West of Ireland' in B. O' Connor and M. Cronin (eds), *Tourism in Ireland: A Critical Analysis* (Cork 1993).

Cabot, D., *The New Naturalist: Ireland* (London 1999).

Cadogan, S., 'Travel Curbs Could Lead to a Dead End', *Irish Examiner Farming Supplement* 21 Aug. 2003.

Calhoun, C., 'Habitus, Field and Capital: The Question of Historical Specificity' in C. Calhoun, E. LiPuma and M. Postone *Bourdieu: Critical Perspectives* (Cambridge 1993).

Caslin, B., 'REPS: A Western Revolution for the Farming Environment', *Farmer's Journal*, 56, 17 (2003).

Cawley, M., 'Poverty and Accessibility to Services in the Rural West of

Bibliography

Ireland' in D.G. Pringle *et al.* (eds), *Poor People, Poor Places: A Geography of Poverty and Deprivation in Ireland* (Dublin 1999).

Central Statistics Office (CSO), *Farming Since the Famine – Irish Farm Statistics 1847–1996* (Cork 1997).

Chayanov, A.V., 'The Theory of Peasant Economy' quoted in B. Kerblay 'Chayanov and the Theory of Peasantry as a Specific Type of Economy' in T. Shanin (ed.), *Peasants and Peasant Societies* (Harmondsworth 1984).

Chomsky, N., *Rogue States: The Rule of Force in World Affairs* (London 2000).

Chubb, B., *The Government and Politics of Ireland* (Dublin 1991).

Clark, J., and P. Lowe, 'Cleaning Up Agriculture: Environment, Technology and Social Science', *Sociologia Ruralis*, 32, 1 (1992), 11–29.

Cloke, P., 'Counter-urbanisation: A Rural Perspective', *Geography*, 70 (1985), 13–23.

Cohen, R., and M. Rai (2000) *Global Social Movements* (London 2000).

Commins, P., 'Restructuring Agriculture in Advanced Societies: Transformation, Crisis and Responses' in T. Marsden, P. Lowe and S. Whatmore (eds), *Rural Restructuring: Global Processes and Their Responses* (London 1990).

Commins, P., 'The European Community and the Irish Rural Economy' in P. Clancy *et al.* (eds), *Irish Society: Sociological Perspectives* (Dublin 1995).

Commins, P., and J.P. Frawley, *The Development of West Cork – Trends, Plans and Prospects* (Dublin 1994).

Commins, P., and M.J. Keane, *Developing the Rural Economy: Problems, Programmes and Prospects* (Dublin 1994).

Connelly, J., and G. Smith, *Politics and the Environment: From Theory to Practice* (London 1999).

Connolly, L., *An Analysis of Farm Structures and Incomes* available at www.teagasc.ie/publications/2002.

Connolly, L., A. Kinsella and G. Quinlan, *National Farm Survey 2002* (Dublin 2003).

Cooney, G., T. Condit and E. Byrnes, 'The Archaeological Landscape' in N. Buttimer, C. Rynne and H. Guerin (eds), *The Heritage of Ireland* (Cork 2000).

Coughlan, A., 'Ireland' in D. Seers and C.Vaitsos (eds), *Integration and Unequal Development: The Experience of the EEC* (London 1980).

Coulter, C., *The Hidden Tradition: Feminism, Women and Nationalism in Ireland* (Cork 1993).

Cox, P.G., 'The Impact of EC Structural Policy in Ireland' in An Foras Taluntais, *The Challenge Facing Agriculture in Difficult Times* (Dublin 1985).

Crowley, E., 'Making A Difference?: Female Employment And Multinationals In The Republic Of Ireland' in M. Leonard and A. Byrne (eds),

Bibliography

Women and Ireland: A Sociological Reader (Belfast 1997).

Crowley, E., 'Towards Sustainable Agriculture?: A Sociological Analysis of The Rural Environment Protection Scheme (REPS) in the South-West of Ireland' (Unpublished PhD dissertation, NUI, Cork 2000).

Crowley, E., 'The Evolution of the Common Agricultural Policy and Social Differentiation in Rural Ireland', *The Economic and Social Review*, 23, 1 (2003).

Crowley, E., 'Local Exchange Trading Systems: Globalizing Rural Communities' (Discussion Papers Series, Institute for International Integration Studies (IIIS), Trinity College Dublin 2004).

Culleton, N., and P. Dillon, 'Why Ireland Can Justify a Higher Nitrates Limit', *Farmers Journal*, 57, 27 (2004).

Cypher, J.M., and J.L. Dietz, *The Process of Economic Development* (London 1997).

Daniels, S., and D. Cosgrove, 'Introduction' in D. Cosgrove and S. Daniels (eds), *The Iconography of Landscape* (Cambridge 1988).

Della Porta, D., H. Kriesi and D. Rucht, *Social Movements in a Globalising World* (Basingstoke 1999).

Dept. of Agriculture and Food (DAF), *Towards a Sustainable Land Policy* (Dublin 1997).

Deverre, C., 'Social Implications of Agro-Environmental Policy in France and Europe', *Sociologia Ruralis*, 35, 2 (1995), 227–47.

Devlin, J., 'Guidelines to Local Authorities on Sustainable Rural Housing – the IFA Position', *Farmer's Journal*, 57, 27 (2004).

Dickson, D., 'Butter Comes to the Market: The Origins of Commercial Dairying in County Cork' in P. O'Flanagan and C.G. Buttimer (eds), *Cork: History and Society* (Dublin 1993).

DiMaggio, P., 'Review Essay: On Pierre Bourdieu', *American Journal of Sociology*, 84, 6 (1979), 1460–74.

Dineen, M., 'REPS Chaos in the West Angers Farmers', *Farming Independent* 26 Aug. 2003.

Donnelly, J.S., *The Land and People of Nineteenth-Century Cork* (London 1975).

Donnelly, M., 'Very Small Amount of Land Is Put on the Market, Says CSO', *Farming Independent* 26 Aug. 2003.

Donnelly, M., 'Farm Nutrient Plans Would Combat Pollution', *Farming Independent* 26 Aug. 2003.

Donnelly, W.J., and S. Crosse, 'A Perspective for the Dairy Industry 2010', *Farm and Food*, 10, 1 (2000), 4–18.

Dooney, S., *Irish Agriculture: An Organisational Profile* (Dublin 1988).

Douthwaite, R., *Short Circuit: Strengthening Local Economies for Security in an Unstable World* (Dublin 1996).

Eder, K., 'The Institutionalisation of Environmentalism: Ecological Discourse and the Second Transformation of the Public Sphere' in S. Lash, B. Szerszynski and B. Wynne (eds), *Risk, Environment and*

Bibliography

Modernity: Towards A New Ecology (London 1996).

Eipper, C., *The Ruling Trinity: A Community Study of Church, State and Business in Ireland* (Aldershot 1986).

Eipper, C., *Hostage to Fortune – Bantry Bay and the Encounter with Gulf Oil* (Newfoundland 1989).

ENFO, 'Fish Kills in Ireland' Fact Sheet 3 (Dublin 2000).

Escobar, A., 'Constructing Nature: Elements for a Poststructural Political Ecology' in R. Peet and M. Watts (eds), *Liberation Ecologies: Environment, Development, Social Movements* (London 1996).

European Commission, 'A New Start For the Common Agricultural Policy' *Green Europe Newsletter* (Dec. 1980).

European Commission, *The Agricultural Situation in the European Community – 1984 Report* (1985).

European Commission, *The Agricultural Situation in the European Community – 1986 Report* (1987).

European Commission, *Intensive Farming and the Impact on the Environment and the Rural Economy of Restrictions on the Use of Chemical and Animal Fertilisers* (1989).

European Commission, 'Council Regulation 2078/92' in *Official Journal of the European Communities* No.L, 215 (July 1992).

European Commission, *Taking European Environment Policy into the 21st Century* (1996).

European Commission, *The Agricultural Situation in the European Community – 1996 Report* (1997).

EUROSTAT Yearbook, *A Statistical Eye on Europe 1986–1996* (Luxembourg 1997).

Evans, E.E., *Irish Folk Ways* (London 1957).

Evans, M., and L. Coen, 'Elitism and Agri-Environmental Policy in Ireland' in M. Adshead and M. Millar (eds), *Public Administration and Public Policy in Ireland: Theory and Methods* (London 2001).

Feehan, J., 'Threat and Conservation: Attitudes to Nature in Ireland' in J.W. Foster and H. Chesney (eds), *Nature In Ireland: A Scientific and Cultural History* (Dublin 1997).

Feehan, J., *Farming in Ireland: History, Heritage and Environment* (Dublin 2003).

Finn, J.A., 'Agri-Environmental Policy and Environmental Quality', *Farm and Food*, 13, 1 (2003).

Fischler, F., 'CAP 2000 – A Framework for the Future of EU Agriculture', *Farmer's Journal*, 49, 36 (1997).

Forsythe, D., 'Urban Incomers and Rural Change – The Impact of Migrants from the City on Life in an Orkney Community', *Sociologia Ruralis*, 20, 4 (1980), 287–307.

Frawley, J., and P. Commins, *The Changing Structure of Irish Farming* (Dublin 1996).

Friemann, G., 'Looking For Organic Growth', *Irish Times* 20 Mar. 2004.

Bibliography

Fröbel, F.J. *et al.*, *The New International Division of Labour* (Cambridge 1980).

Frouws, J., 'The Contested Redefinition of the Countryside: An Analysis of Rural Discourses in the Netherlands', *Sociologia Ruralis*, 38, 1 (1998), 54–68.

Fuchsia Brands Ltd., www.fuchsiabrands.com (2001)

Garland, R., 'Problems Regarding Access to the Irish Landscape' in T. O' Regan (ed.), *Through the Eye of the Artist* (Maynooth 1998).

Giddens, A., *The Consequences of Modernity* (Cambridge 1990).

Goodman, D., and M. Redclift, *From Peasant to Proletarian: Capitalist Development and Agrarian Transitions* (Oxford 1981).

Goodman, D., and M. Watts, 'Reconfiguring the Rural or Fording the Divide?: Capitalist Restructuring and the Global Agro-Food System', *Journal of Peasant Studies*, 22, 1 (1994), 1–49.

Green Group, *Farming With Nature* (Brussels 1993).

Gribaudi, G., 'Images of the South' in D. Forgacs and R. Lumley (eds), *Italian Cultural Studies – An Introduction* (Oxford 1996).

Hajer, M., 'Ecological Modernisation as Cultural Politics' in S. Lash, B. Szerszynski and B. Wynne (eds), *Risk, Environment and Modernity: Towards A New Ecology* (London 1996).

Halfacree, K., 'Contrasting Roles for the Post-Productivist Countryside' in P. Cloke and J. Little (eds), *Contested Countryside Cultures: Otherness, Marginalisation, Rurality* (London 1997).

Hannan, D.F., *Displacement and Development: Class, Kinship and Social Change in Irish Rural Communities* (Dublin 1979).

Hannan, D.F., and P. Commins, 'The Significance of Small-Scale Landholders in Ireland's Socio-economic Transformation' in J.H. Goldthorpe and C.T. Whelan (eds), *The Development of Industrial Society in Ireland* (New York 1992).

Hannigan, J.A., *Environmental Sociology: A Social Constructionist Perspective* (London 1995).

Harker, R. *et al.* (eds), *An Introduction to the Work of Pierre Bourdieu* (Basingstoke 1990), especially 'Conclusion: Critique'.

Harvey, D., *The Condition of Postmodernity* (Massachusetts 1989).

Harvey, D., *Justice, Nature and the Geography of Difference* (Massachusetts 1996).

Harvey, L., and M. MacDonald, *Doing Sociology: A Practical Introduction* (London 1993).

Haynes, A., 'Social Capital and Regional Growth' in M. Peillon and M.P. Corcoran (eds), *Place and Non-Place: The Reconfiguration of Ireland* (Dublin 2004).

Heaney, S., *The Government of the Tongue* (Harmondsworth 1990).

Hegarty, H., 'A Geographical Analysis of the Socio-Cultural Interface Between the Locals and the Incomers in West Cork' (Unpublished MA Thesis, NUI, Cork 1994).

Bibliography

Hendrickson, M., and W.D. Heffernan, 'Opening Spaces Through Relocalization: Locating Potential Resistance in the Weaknesses of the Global Food System', *Sociologia Ruralis*, 42, 4 (2002), 347–69.

Herman, D., 'A Land of Few Welcomes for Visiting Hill Walkers', *Irish Times* 10 Jan. 2004.

Hetherington, K., 'Stonehenge and its Festival: Spaces of Consumption' in R. Shields (ed.), *Lifestyle Shopping: The Subject of Consumption* (London 1992).

Hines, C., *Localization: A Global Manifesto* (London 2000).

Ilbury, B. *et al.*, 'Quality, Imagery and Marketing: Producer Perspectives on Quality Products and Services in the Lagging Rural Regions of the European Union', *Geografiska Annaler Series B*, 83, 1 (2001), 27–40.

Jackson, J., and T. Haase, 'Demography and the Distribution of Deprivation in Rural Ireland' in C. Curtin, T. Haase and H. Tovey (eds), *Poverty in Rural Ireland: A Political Economy Perspective* (Dublin 1996).

Jameson, F., 'Postmodernism, or, the Cultural Logic of Late Capitalism', *New Left Review*, 146 (1984), 59–92.

Kalb, D., 'Localizing Flows: Power, Paths, Institutions and Networks' in D. Kalb *et al.* (eds), *The Ends of Globalization: Bringing Society Back In* (Lanham 2000).

Keohane, K., 'Reflexive Modernisation and Systematically Distorted Communications: An Analysis of an Environmental Protection Agency Hearing', *Irish Journal of Sociology*, 8 (1998), 71–92.

Kerblay, B., 'Chayanov and the Theory of Peasantry as a Specific Type of Economy' in T. Shanin (ed.), *Peasants and Peasant Societies* (Harmondsworth 1984).

Kinsella, J. *et al.*, 'Pluriactivity as a Livelihood Strategy in Irish Farm Households and its Role in Rural Development', *Sociologia Ruralis*, 40, 4 (2000), 481–96.

Lalor, B., *West of West: An Artist's Encounter with West Cork* (Dingle 1990).

Lavery, M., 'Landowners Up in Arms', *Farmer's Journal*, 57, 22 (2004).

Lechner, F.J., and J. Boli, *The Globalisation Reader* (Blackwell 2000).

Lee, D.J. and B.S. Turner, *Conflicts About Class: Debating Inequality in Late Industrialism* (London 1996).

Lee, J., 'Environmental Indicators and Sustainable Agriculture', *Farm and Food*, 9, 1 (1999).

Lee, R., 'Moral Money? LETS and the Social Construction of Local Economic Geographies in Southeast England', *Environment and Planning, A*, 28 (1996), 1377–94.

Leeuwis, C., *Marginalisation Misunderstood* (Wageningen 1989).

Long, N., 'Commoditization: Thesis and Antithesis' in N. Long *et al.* (eds), *The Commoditization Debate: Labour Process, Strategy and Social Network* (Wageningen 1986).

Long, N., 'Introduction' in N. Long and A. Long (eds), *Battlefields of Knowledge: The Interlocking of Theory and Practice in Social*

Bibliography

Research and Development (London 1992).

Lowe, P., 'Industrial Agriculture and Environmental Regulation: A new Agenda for Rural Sociology', *Sociologia Ruralis*, 32, 1 (1992), 4–10.

Mac Connell, S., 'EU to Fund Access for Walkers in Co. Wicklow', *Irish Times* 14 July 2004.

Mac Connell, S., 'Farmers Warned They Could Lose EU Payment' *Irish Times* 11 Sept. 2004.

McDonald, F., 'We're On the Road to Nowhere', *Irish Times* 14 July 2001.

McDonald, F., 'Planners Despair as Profit Drives New Housing Policy', *Irish Times* 13 March 2004.

McDonald, F., 'Planning to Soldier On Against the Odds', *Irish Times* 20 March 2004(a).

McDonald, F., 'Rural Housing Advocate Advises Tourists to go to Scotland', *Irish Times* 13 March 2004(b).

McDonald, F., 'Running Up a Big Bill for our Rotten Track Record', *Irish Times* 17 July 2004(c).

McDonald, F., 'Tall Tales and Grim Realities', *Irish Times* 5 Sept. 2005.

McEvoy, O., *Impact of REPS – Analysis from the National Farm Survey* (Dublin 1999).

McGrath, B., 'Environmentalism and Property Rights: the Mullaghmore Interpretive Centre Dispute', *Irish Journal of Sociology*, 6 (1996), 25–47.

Mac Laughlin, J., *Ireland: The Emigrant Nursery and the World Economy* (Cork 1994).

Mac Laughlin, J., 'The Devaluation of Nation as Home and the Depoliticisation of Recent Irish Emigration' in J. Mac Laughlin (ed.), *Location and Dislocation in Contemporary Irish Society: Emigration and Irish Identities* (Cork 1997).

Mac Laughlin, J., 'Emigration and the Construction of Nationalist Hegemony in Ireland' in J. Mac Laughlin (ed.), *op. cit.* (1997a)

Mac Laughlin, J., 'The New Vanishing Irish: Social Characteristics of 'New Wave' Irish Emigration' in J. Mac Laughlin (ed.), *op. cit.* (1997b)

McMichael, P., 'Globalisation: Myths and Realities', *Rural Sociology*, 61, 1 (1996), 25–55.

McMichael, P., 'The Power of Food', *Agriculture and Human Values*, 17 (2000), 21–33.

McNally, F., 'Barrier Erected At Disputed Right-of-way', *Irish Times* 20 Sept. 2004.

Maffesoli, M., *The Time of the Tribes* (Cambridge 1996).

Maguire, D., 'SAC Awareness in Wexford', *Journal Plus*, 49, 45 (1997).

Marsden, T. *et al.*, *Constructing the Countryside* (London 1993).

Marsh, J., 'Options for Policies' in J. Marsh *et al.* (eds), *The Changing Role of the Common Agricultural Policy: The Future of Farming in Europe* (London 1991).

Bibliography

Martin, S., 'Democratizing Rural Development' in N. Walford, J. Everitt and D. Napton (eds), *Reshaping the Countryside: Perceptions and Processes of Rural Change* (London 1999).

Marx, K., 'Excerpts from the Eighteenth Brumaire of Louis Bonaparte' in L.S. Feuer (ed.), *Karl Marx & Friedrich Engels: Basic Writings on Politics and Philosophy* (Aylesbury 1984).

Massey, D., *Space, Place and Gender* (Cambridge 1994).

Matthews, A., *Farm Incomes: Myths and Reality* (Cork 2000).

Melucci, A., 'Social Movements and the Democratization of Everyday Life' in J. Keane (ed.), *Civil Society and the State: New European Perspectives* (London 1988).

Midgeley, J., *Social Welfare in Global Context* (California 1997).

Mills, C.W., *The Sociological Imagination* (Oxford 1959).

Mol, A.P.J., 'Ecological Modernisation and Institutional Reflexivity: Environmental Reform in the Late Modern Age', *Environmental Politics*, 5, 2 (1996), 302–23.

Monbiot, G., *Captive State: The Corporate Takeover of Britain* (London 2000).

Monbiot, G., 'The Fruits of Poverty', *The Guardian* 16 March 2004.

Mooney, P., 'SAC Rules Being Revisited', *Farmers Journal*, 49, 38 (1997).

Mooncy, P., 'Big Rowback on SAC/REPS Package', *Farmers Journal*, 50, 8 (1998).

Mooney, P., 'Farmers Plan to Boycott Teagasc', *Farmers Journal*, 57, 23 (2004).

Mormont, M., 'Who is Rural? or, How to be Rural? Towards a Sociology of the Rural' in T. Marsden, P. Lowe, and S. Whatmore (eds), *Rural Restructuring: Global Processes and Their Responses* (London 1990).

Mullane, F., 'Buildings Within the Spirit of the Time' in T. O' Regan (ed.), *Through the Eye of the Artist* (Maynooth 1998).

Murphy, E., and B. Lally, 'Agriculture and Environment in Ireland: Directions for the Future', *Administration*, 46, 1 (1998), 71–98.

Newby, H., 'The Rural Sociology of Advanced Capitalist Societies' in H. Newby (ed.), *International Perspectives in Rural Sociology* (Chichester 1978).

Nolan, P.J., 'Fewer than 25, 000 Commercial Farmers by 2015', *Farmers Journal*, 50, 47 (1998).

Nugent, A., 'Sucklers and Sheep Within REPS', *Farmers Journal*, 52, 29 (2000).

O'Donnell, M., *Introduction to Sociology*, 4th edn (London 1997).

O'Dwyer, J.G., 'Pristine Landscapes Can be Dull Landscapes', *Irish Times* 1 April 2004.

Ó Gráda, C., 'Seasonal Migration and Post-Famine Adjustment in the West of Ireland', *Studia Hibernia* 33 (1973).

O'Hara, P., 'CAP Structural Policy – A New Approach to an Old Problem?' in Economics and Rural Welfare Research Centre *The Changing CAP and its Implications* (Dublin 1986).

O'Hara, P., *Partners in Production? Women, Farm and Family in Ireland* (Oxford 1998*)*

O'Keeffe, P., 'Dairy Farms Now Number 25, 000', *Farmers Journal,* 57, 3 (2004).

Ó Neachtain, P., 'Hill Walking – A Bone of Contention', *Farmers Journal,* 57, 3 (2004).

O'Rielly, S., 'Small and Medium-Sized Food Enterprise Networks: Process and Performance' (Unpublished PhD Dissertation, University of Wales, Aberystwyth 2003).

O'Sullivan, K., 'EPA Report Finds Unrelenting Decline in Freshwater', *Irish Times* 20 May 1999.

Offe, C., *Contradictions of the Welfare State* (London 1984).

Offe, C., and R.G. Heinze, *Beyond Employment: Time, Work and the Informal Economy* (Cambridge 1992).

Orme, A.R., *The World's Landscapes: Ireland* (London 1970).

Ortiz, S., 'Reflections on the Concept of Peasant Culture and Peasant Cognitive Systems' in T. Shanin (ed.), *Peasants and Peasant Societies* (Harmondsworth 1984).

Palmer, J., *Trading Places: The Future of the European Community* (London 1988).

Peace, A., 'A Different Place Altogether': Diversity, Unity, and Boundary in an Irish Village' in A.P. Cohen (ed.), *Symbolising Boundaries: Identity and Diversity in British Cultures* (Manchester 1986).

Philips, P., *Wheat, Europe and the GATT* (London 1990).

Phillips, S.K., 'Natives and Incomers: The Symbolism of Belonging in Muker Parish, North Yorkshire' in A.P. Cohen (ed.), *Symbolising Boundaries: Identity and Diversity in British Cultures* (Manchester 1986).

Phipps, C., 'Organic food: Which is best?', *The Guardian* 10 Dec. 1999.

Pocock, I., 'Slurry Hits the Fan', *Irish Times* 22 May 2004.

Pocock, I., 'State Loses Water Pollution Water Case', *Irish Times* 13 March 2004(a).

Portela, J., 'Agriculture is Primarily What?' in D. Symes and A.J. Jansen (eds), *Agricultural Restructuring and Rural Change in Europe* (Wageningen 1994).

Potter, C., 'Conserving Nature, Agri-Environmental Policy Development and Change' in B. Ilbury (ed.), *The Geography of Rural Change* (Essex 1998).

Purdue, D. *et al.*, 'DIY Culture and Extended Milieux: LETS, Veggie Boxes and Festivals', *The Sociological Review,* 45, 4 (1997).

Quinn, A., *Patrick Kavanagh: A Biography* (Dublin 2003).

Redclift, M., *Sustainable Development: Exploring the Contradictions* (London 1987).

Redclift, M., 'Sustainable Development and Popular Participation: A

Framework for Analysis' in D. Ghai and J.M. Vivian (eds), *Grassroots Environmental Action* (London 1992).

Reid, L., 'EU to Check Alleged Nature Reserve Abuses', *Sunday Tribune* 7 June 1998.

Robertson, R., *Globalisation* (London 1992).

Sage, C., 'Social Embeddedness and Relations of Regard: Alternative 'Good Food' Networks in South-West Ireland', *Journal of Rural Studies,* 19 (2003), 47–60.

Schmalenbach, H., *On Society and Experience: Selected Papers* (Chicago 1977), translated by G. Lüschen and G.P. Stone.

Schulte, R. *et al.*, 'Animal Welfare – Development of Methodology for its Assessment', *Farm and Food,* 8, 3 (1998), 20–3.

Scott, J., *Development Dilemmas in the European Community: Rethinking Regional Development Policy* (Buckingham 1995).

Sheridan, K., 'When Town and Country Collide', *Irish Times* 3 July 2004.

Shiel, T., 'Mobile Farm to Be Class Act in Connacht', *Irish Times* 11 March 2003.

Shirley, J., 'Destocking Shocks in New REPS Scheme', *Farmers Journal,* 50, 28, 1998.

Shucksmith, M., 'Territorial Aspects of the Common Agricultural Policy' Paper delivered to 2003 European Congress of Rural Sociology, Sligo, Ireland (2003).

Shutes, M.T., 'Kerry Farmers and the European Community: Capital Transitions in a Rural Irish Parish', *Irish Journal of Sociology,* 1 (1991), 1–17.

Slater, E., 'Cast a Cold Eye: Packaging Irish Heritage', *Irish Reporter,* 10 Second Quarter (1993), 8–11.

Smith, M., 'One-off Housing is Bad for Society and the Environment', *Irish Times* 30 May 2004.

Staunton, D., 'State in Breach of EU Environmental Law', *Irish Times* 14 July 2004.

Storey, D., 'The Spatial Distribution of Education and Health and Welfare Facilities in Rural Ireland', *Administration,* 42, 3 (Autumn 1994), 246–68.

Storey, D., 'Issues of Integration, Participation and Empowerment in Rural Development: The Case of LEADER in the Republic of Ireland', *Journal of Rural Studies,* 15, 3 (1999), 307–15.

Strassoldo, R., 'Globalism and Localism: Theoretical Reflections and Some Evidence' in Z. Mlinar (ed.), *Globalisation and Territorial Identities* (Aldershot 1992).

Symes, D., 'Agriculture, the State and Rural Society in Europe: Trends and Issues', *Sociologia Ruralis,* 32, 2–3 (1992), 193–208.

Thompson, E.P., *The Making of the English Working Class* (Harmondsworth 1968).

Thorne, L., 'Local Exchange Trading Systems in the United Kingdom: A

Bibliography

case of re-embedding?', *Environment and Planning, A,* 28 (1996), 1361–76.

Tovey, H., 'Milking the Farmer? Modernisation and Marginalisation in Irish Dairy Farming' in M. Kelly, L. O'Dowd and J. Wickham (eds), *Power, Conflict And Inequality* (Dublin 1982).

Tovey, H., 'Rural Management, Public Discourses and the Farmer as Environmental Actor' in D. Symes and A.J. Jansen (eds), *Agricultural Restructuring and Rural Change in Europe* (Wageningen 1994).

Tovey, H., 'Natural Resource Development and Rural Poverty' in C. Curtin, T. Haase and H. Tovey (eds), *Poverty in Rural Ireland: A Political Economy Perspective* (Dublin 1996).

Tovey, H., ' "We Can All Use Calculators Now": Productionism, Sustainability, and the Professional Formation of Farming in Co. Meath, Ireland' in H. de Haan, B. Kasimis and M. Redclift (eds), *Sustainable Rural Development* (Aldershot 1997).

Tovey, H., 'Food, Environmentalism and Rural Sociology: On the Organic Farming Movement in Ireland', *Sociologia Ruralis,* 37, 1 (1997), 21–37.

Tovey, H., ' "Messers, Visionaries and Organobureaucrats": Dilemmas of Institutionalisation in the Irish Organic Farming Movement', *Irish Journal of Sociology,* 9 (1999), 31–59.

Tovey, H., T. Haase and C. Curtin, 'Understanding Rural Poverty' in C. Curtin, T. Haase and H. Tovey (eds), *Poverty in Rural Ireland: A Political Economy Perspective* (Dublin 1996).

Tovey, H., and P. Share, *A Sociology of Ireland,* 2nd edn (Dublin 2003).

Urry, J., *The Tourist Gaze* (London 1990).

Urry, J., *Consuming Places* (London 1995).

Vail, D., K.P. Hasund and L. Drake, *The Greening of Agricultural Policy in Industrial Societies* (Ithaca 1994).

van der Ploeg, J.D., 'Styles of Farming: An Introductory Note on Concepts and Methodology' in J.D. van der Ploeg, and A. Long (eds), *Born From Within: Practice and Perspectives of Endogenous Rural Development* (Assen 1994).

Viney, M., *A Living Island: Ireland's Responsibility to Nature* (Dublin 2003).

Wacquant, L.J.D., 'Toward a Social Praxeology: The Structure and Logic of Bourdieu's Sociology' in P. Bourdieu and L.J.D. Wacquant, *An Invitation to Reflexive Sociology* (Cambridge 1992).

Wacquant, L.J.D., 'Pierre Bourdieu' in R. Stones (ed.) *Key Sociological Thinkers* (London 1998).

Wallace, R.A., and A. Wolf (eds), *Contemporary Sociological Theory: Continuing the Classical Tradition* (Englewood Cliffs 1995).

Walsh, D.F., 'Structure/Agency' in C. Jenks (ed.), *Core Sociological Dichotomies* (London 1998).

Ward, N., 'The Agricultural Treadmill and the Rural Environment in the

Bibliography

Post-Productivist Era', *Sociologia Ruralis*, 33, 3–4 (1993), 348–64.

WCED, *Our Common Future – The Brundtland Report* (Oxford 1987).

West Cork LEADER Ltd (n.d.) www.westcorkleader.ie

Whatmore, S., 'Theories and Practices for Rural Sociology in a 'New' Europe', *Sociologia Ruralis*, 30, 3–4 (1990), 251–9.

Whelan, K., 'The Bases of Regionalism' in P. Ó Drisceoil (ed.), *Culture in Ireland – Regions: Identity and Power* (Belfast 1993).

Whitby, M., 'The Prospect for Agri-Environmental Policies within a Reformed CAP' in M. Whitby (ed.), *The European Environment and CAP Reform: Policies and Prospects for Conservation* (Wallingford 1996).

Williams, C.C., 'Local Exchange and Trading Systems: A new source of work and credit for the poor and unemployed?', *Environment and Planning A*, 28 (1996), 1395–1415.

Wilson, T.M., 'Beef Before the Ballot: The Impact of the Common Market on Agriculture and Politics in Eastern Ireland' (Unpublished PhD dissertation, City University of New York, New York 1985).

Wilson, T.M., 'Large Farms, Local Politics, and the International Arena: The Irish Tax Dispute of 1979', *Human Organisation*, 48, 1 (1989), 60–70.

Young, P., 'Part-time Farming: Friend or Foe of Irish Agriculture?', *Farmers Journal*, 56, 32 (2003).

Select Index

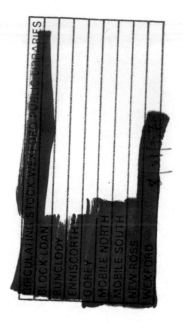